Contents

Preface	iv
Assessment guide	iv
Experiments	1
1. Looking at cells	1
2. Plant tissue culture	9
3. Tryptic digest of casein	14
4. Extracting DNA from onions	20
5. Food testing	23
6. The chromatography of plant pigments	27
7. The effect of sunlight on photosynthesis	32
8. Aerobic and anaerobic respiration in yeast	37
9. Questionnaire on the Human Genome Project	41
10. Looking at DNA fingerprints	44
11. The extraction of polyphenol oxidase from mushrooms	52
12. The action of pepsin on egg white	57
13. Investigating the thermal stability of chymosin	60
14. The effect of pH on enzyme activity	64
15. The effect of substrate concentration on enzyme action	68
16. Designing a washing powder	72
17. Investigating enzyme specificity using molecular modelling	76
18. Chemicals and cancer (a data-handling exercise)	81
List of suppliers	89
Bibliography	89

Preface

This practical guide accompanies the book *Biochemistry for advanced biology* (ISBN 0521 437814). There are 18 experiments, which mainly follow the order in which the topics are introduced in the text.

A feature of this guide is the setting of the experiments into a larger context as given in the introduction to each. More information is also to be found in *Biochemistry for advanced biology* and a chapter reference is given. The teacher/technician notes list the equipment and materials required, with details of recipes and suppliers as well as answers to the students' notes. Unless otherwise stated, all materials are obtainable from Philip Harris. Other suppliers are listed at the back of the book.

The assessment guide that follows each chapter is derived from the requirements of the A level examination boards.

Assessment guide

Each experiment in this guide may be linked with teacher assessment. The skills that can be assessed fall into the following four categories, which have been formulated from the current common requirements of the A level examination boards.

1. Experimental skills: includes handling apparatus and materials, safety, collecting data, organisation and a reasonable balance between team and independent work in the laboratory.
2. Observation and recording: includes systematic and accurate taking of readings and other observations, recording of data and observations in an acceptable format.
3. Data handling: includes appreciation of limitations of experiment and sources of error, handling of mathematical calculations, drawing graphs, use of statistics; also includes linking data to conclusion from patterns and trends without disregarding any unexpected results.
4. Experimental design: includes production of a detailed and logical written plan of work, with attention paid to the use of controls as well as an appreciation of how the data will be handled to produce a conclusion.

The assessment guide for each experiment focuses on some of the above skills and is a mark scheme. The figures in parentheses are to be used as marks that give a total of ten.

Experiment 1
Looking at cells

Aim
This experiment will give you practice in preparing and staining microscope slides and handling the light microscope. It will also give you experience in identifying the main structures (organelles and membrane systems) of eukaryotic cells from electron micrographs.

Introduction
The largest cells are plant cells, with diameters between 10 and 100 μm. Animal cells have diameters between 10 and 30 μm. Single-celled organisms such as bacteria are much smaller, with diameters of only about 1 μm (you could fit thousands of these cells onto a full stop). This means that cells can only be seen with a microscope.

The development of the German dye industry in the nineteenth century led to the use of many stains in microscopy. Stains are absorbed by important structures inside the cell, such as nuclei and chromosomes, which can therefore be seen more clearly. The use of Gram's stain allows bacteria to be divided into two main classes (**Gram-positive** and **Gram-negative**), which differ in the chemical composition of their cell walls. Gram-positive bacteria include *Pneumococcus* species, some of which cause bacterial pneumonia, while Gram-negatives include *Escherichia coli*, which can be responsible for gastro-enteritis and urinary infections. As Gram-positives are sensitive to penicillin and Gram-negatives are not, the distinction between them is important in medical microbiology.

Looking at the shape (morphology) of cells can also be important clinically. The smear test, which is used to detect cancer of the cervix, involves the collection of a sample of cells from the cervix. Looking at these under the microscope enables changes that may lead to cancer to be detected early enough for effective treatment to be given.

(See also Chapter 1, *Biochemistry for advanced biology*.)

Equipment and materials
- microscope
- oil immersion lens
- cedarwood oil
- microscope slides
- cover slips
- wooden spatula
- tweezers
- scalpel
- dropping pipettes
- bunsen burner
- liquid bacterial culture
- onion, or other plant material
- stains: methylene blue, iodine, crystal violet (Gram's stain), carbol fuchsin
- ethanol
- electron micrographs

© Cambridge University Press 1994

SAFETY

Ethanol is flammable – keep it away from the bunsen burner.

Mop up any spills of *bacterial culture* with ethanol.

Experimental procedure

Animal cells

1. Use the spatula to remove a few cells from the inside of your wrist – you only need to rub gently, as these cells are shed very easily.
2. Smear the spatula onto a slide and add a drop of methylene blue.
3. Cover with cover slip, taking care not to trap air bubbles (see Fig. 1.1).
4. Focus the microscope with the lowest-power objective first. Sketch and label what you see.
5. Look at the slide with the other objectives.

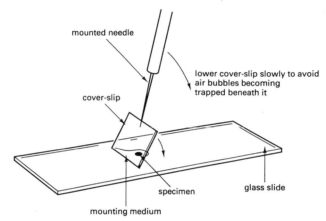

Figure 1.1 Lowering a cover slip to avoid obtaining bubbles below it.

Plant cells

1. With a scalpel, detach a small piece of the thin skin between the layers of an onion.
2. Make a slide as described above, staining with iodine.
3. Examine and record as above.

Bacteria

1. Place one drop of the liquid bacterial culture on a slide.
2. Allow it to dry in the air.
3. Pass the slide three or four times through a bunsen flame.
4. Cover with a few drops of crystal violet and leave for 30 seconds.
5. Wash with water using a dropping pipette.
6. Cover with a few drops of iodine and leave for 30 seconds.
9. Wash with ethanol, until the washings are pale purple.
10. Wash with water.
11. Cover with a few drops of carbol fuchsin for 10 seconds.
12. Wash with water.
13. Gently blot the slide dry with a tissue. Examine under the microscope at a magnification of ×1000 with an oil immersion lens, after gently blotting dry with a tissue. Pink cells are Gram-negative, blue/purple ones are Gram-positive. Record your observations.

© Cambridge University Press 1994

Electron micrographs

1. Identify the nucleus, nuclear envelope, endoplasmic reticulum, mitochondria and plasma membrane in the rat pancreas exocrine cell, below. (Figure 1.2a).

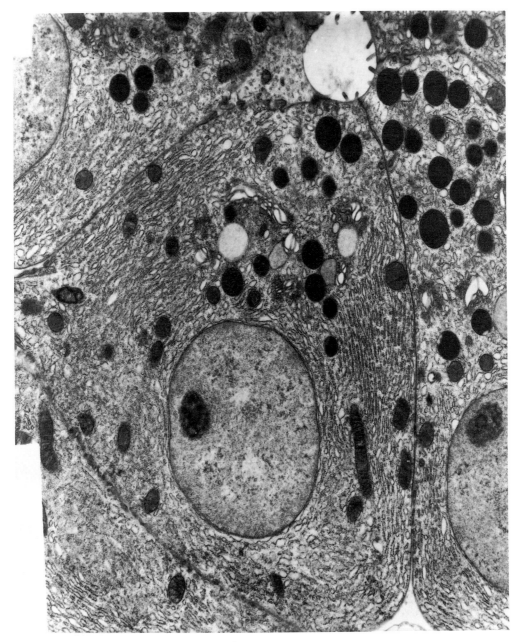

Figure 1.2 (a) Electron micrograph (unlabelled) of a thin section of a representative animal cell, a rat pancreas exocrine cell ×10 000).

2. Identify as many organelles as you can in the leaf mesophyll cell (Figure 1.2b), including the chloroplasts.

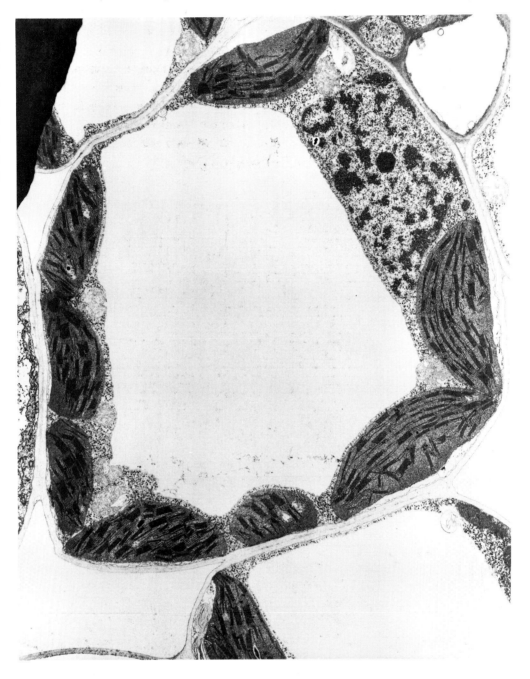

(b) Electron micrograph (unlabelled) of a thin section of a representative plant cell, a leaf mesophyll cell (×13 000).

© Cambridge University Press 1994

QUESTIONS

1. Use your observations to summarise the differences between prokaryotic and eukaryotic cells, and between plant and animal cells.
2. It is sometimes important to count numbers of cells (e.g. blood cells) in investigating diseases such as anaemia, and bacteria in public health work. This is done with a device known as a haemocytometer, which is a modified microscope slide with a depression 0.1 mm deep and a grid of total area 1.00 mm^2, as shown in Figure 1.3. If the average number of cells on each of the second-order squares (one is blacked out in a larger, shaded square) is three, what is the number of cells per cubic centimetre (cm^3)?
3. Figure 1.4 is drawn from an electron micrograph of a cell. Calculate the magnification of the diagram. Show your working.

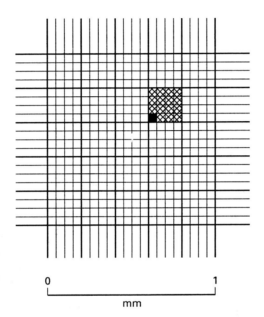

depth of sample over grid = 0.1 mm

Figure 1.3 Haemocytometer grid.

Figure 1.4.

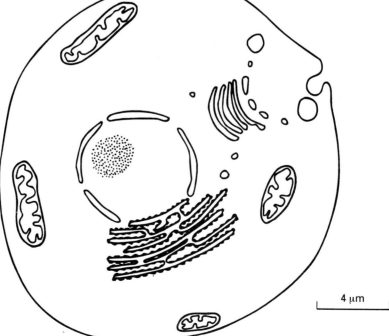

4 μm

Teacher/technician notes

Equipment and materials
Each student or group will need:

- access to microscope
- oil immersion lens – needed to examine bacterial culture
- cedarwood oil
- microscope slides and cover slips
- bunsen burner
- liquid bacterial culture (*Bacillus subtilis* is Gram-positive, while *Azotobacter* species are Gram-negative; cultures of both are obtained by inoculating $100\,cm^3$ nutrient broth with a commercial culture and incubating for 48 hours at $30\,°C$)
- wooden spatula
- tweezers
- scalpel
- dropping pipettes
- onion, or other plant material
- stains: methylene blue (0.125 per cent methylene blue in 0.75 per cent sodium chloride solution); iodine (1 g iodine, 2 g potassium iodide in $300\,cm^3$ distilled water); crystal violet (Gram's stain) (0.5 per cent aqueous solution); fuchsin (1 per cent aqueous solution)
- ethanol
- electron micrographs (Figures 1.2a, b, and 1.5a, b)

Answers to questions

1. See Figures 1.5a and b. Eukaryotes have nuclei, and more organelles. Plants have cell walls, chloroplasts and vacuoles.
2. Area of small square is $0.05 \times 0.05\,mm^2$.
   Volume of small 'square' is $0.05 \times 0.05 \times 0.1\,mm^3$.
   Since there are $10^3\,mm^3$ in $1\,cm^3$, the volume of a small 'square' is $0.05 \times 0.05 \times 0.1 \times 10^{-3} = 2.5 \times 10^{-7}\,cm^3$.
   There are $1/(2.5 \times 10^{-7})$ or 4×10^6 small squares in $1\,cm^3$.
   If each one contains an average of three cells, the approximate number of cells per cm^3 is 1.2×10^7.
3. 2 cm is equivalent to $4\,\mu m$. 1 cm is equivalent to $2\,\mu m$
   1 cm equals $10\,000\,\mu m$.
   Magnification is 5000 (check, $4\,\mu m \times 5000 = 20\,000\,\mu m$ or 2 cm).

Assessment guide
Observation and recording

1. Clear, detailed line drawings of cells (4).
2. Labels, including magnification (4).
3. Correct assignment of features in electron micrographs (2).

© Cambridge University Press 1994

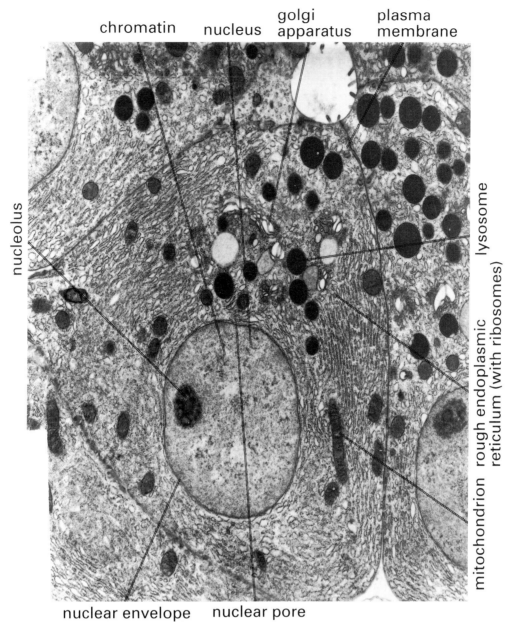

Figure 1.5 (a) Electron micrograph (labelled) of a thin section of a representative animal cell, a rat pancreas exocrine cell (×**10 000**).

© Cambridge University Press 1994

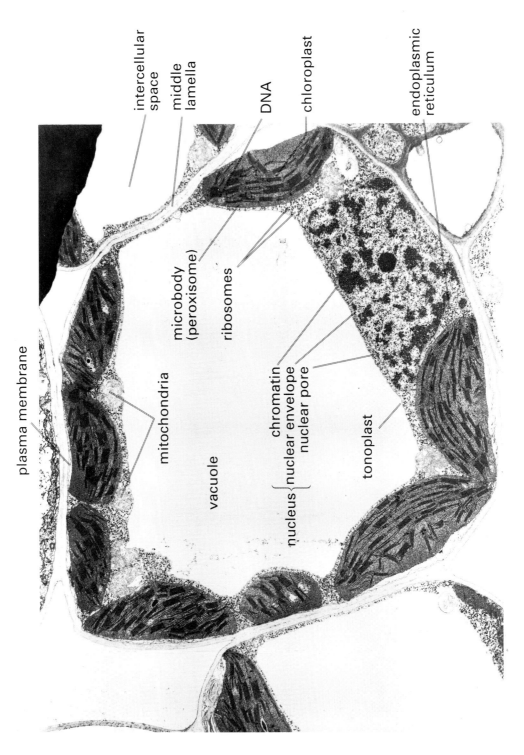

(b) Electron micrograph (labelled) of a thin section of a representative plant cell, a leaf mesophyll cell ($\times$12 000).

Experiment 2
Plant tissue culture

Aim
In this investigation you will observe examples of **totipotency** – the ability of a plant to recreate the whole organism from a single cell, and the action of plant hormones in controlling the growth and form of a plant.

Introduction
It is possible to regenerate a whole plant from a single plant cell. This means that many genetically identical plants – known as **clones** – can be produced from plants with desirable agricultural characteristics. First, a group of cells is isolated from the plant and sterilised. These can come from any part of the plant. These are then grown in a specially formulated nutrient medium. The addition of plant hormones to the medium allows the development of roots and shoots. The tiny plants (plantlets) are then grown on in greenhouses, and then out of doors.

Growing plants in this way promises to have a significant impact on world agriculture. In Vietnam, farmers already use tissue culture to grow disease-free potatoes, and in Thailand there is interest in growing orchid clones as a cash crop.

(See also Chapter 1, *Biochemistry for advanced biology*.)

Equipment and materials
- sterile Petri dishes
- tweezers and scalpel
- cuttings from a rose bush
- cuttings from sundew plant
- cauliflower floret
- barley seeds
- sodium hypochlorite sterilising solution
- sterile distilled water
- M & S (Murashige and Skoog) medium – with various plant hormone additives in containers
- ethanol

Note on plant hormones. You use two types of plant hormone in this experiment – **auxins** and **cytokinins**. The auxins are IAA, 2,4-D and 2-NAA and the cytokinins are kinetin and BAP. A high cytokinin to auxin ratio generally stimulates shoot growth; a low cytokinin to auxin ratio favours roots. A mixture containing roughly equal quantitites tends to produce an undifferentiated mass of tissue known as **callus**.

SAFETY

Keep bunsen burner well away from *ethanol* solution.

© Cambridge University Press 1994

Experimental procedure

Barley
Dissect out two or three embryos from barley seeds (practice this first to locate the embryos within the seeds). Cover them with sodium hypochlorite for 10 minutes in a Petri dish to sterilise them. Then rinse with sterile distilled water several times. Place aseptically on the solid medium provided (this is basic M & S with agar to solidify it, 8 mg per litre of IAA and 0.25 mg per litre of kinetin). Leave for several weeks in a warm place, exposed to the light. Examine weekly for signs of growth.

Rose
Take a terminal bud cutting and remove outer leaves to give about 1.5 cm length cutting. Sterilise as before, cut off the base and place in rose medium. This is the same as the barley medium above, except that the hormone additives are 2 mg per litre of 6-benzylaminopurine (BAP) and 0.01 mg per litre of 1-naphthylacetic acid (NAA).

Cauliflower
Take two or three pieces of cauliflower florets about half a centimetre square. Sterilise and place onto cauliflower medium. This is basically the same as the barley medium, but the agar is omitted to keep it liquid.

Sundew
Cut small clumps of tissue from the centre of a sundew plant. Sterilise and place on sundew medium. This has the same composition as the barley medium.

QUESTIONS
1. Make a table that describes the appearance of each plant over a period of several weeks. Each week, enter a description (with the date), which should include the appearance of shoots, roots, stem and leaves.
2. Why must the medium and the plant material be sterilised?
3. Look at Table 2.1, which is the recipe for M & S medium. What are the functions of the various ingredients?
4. What other crops may benefit from propagation by tissue culture?
5. 2,4-D is a plant hormone with a similar effect to IAA. It is used to kill weeds in lawns and in cereal crops. Can you suggest how it works?

Teacher/technician notes

Equipment and materials

- sterile Petri dishes
- tweezers and scalpel
- cuttings from a rose bush (the younger the bush, the more quickly growth will be established)
- cauliflower floret
- barley seeds
- sundew plant (point out to students that this is a carnivorous plant that feeds on insects)
- sodium hypochlorite solution (10 per cent in water)
- ethanol
- sterile distilled water
- various media as decribed below

Table 2.1 The composition of M & S medium

ammonium nitrate	zinc sulphate
calcium chloride	disodium ethylenediamine tetra-acetate
magnesium sulphate	iron sulphate
potassium dihydrogenphosphate	glycine
potassium nitrate	meso-inositol
boric acid	nicotinic acid
cobalt chloride	pyridoxine hydrochloride
copper sulphate	thiamine hydrochloride
manganese sulphate	sucrose
potassium iodide	indol-3-yl acetic acid (IAA)
sodium molybdate	kinetin

M & S medium

The basic M & S is available from Philip Harris, but if you want to make up your own, quantities are given below, in the solution to question 3. The pH is adjusted to 5.8 with 0.2 M potassium hydroxide before adding the agar, if used. It should then be sterilised at 121 °C/15 psi for 20 minutes. The media for each plant are as follows.

Barley medium

Basic M & S with 9 g per litre of agar to solidify, IAA adjusted to 8 mg per litre and kinetin to 0.25 mg per litre. Pour into individual plastic pots and label, indicating hormone additive. Students may question the large excess of IAA – this is because it breaks down very easily in the plant, so extra is added to compensate for this.

Cauliflower medium

As barley but without the agar to obtain a liquid medium, which can be dispensed into small conical flasks and labelled as above.

Rose medium

Basic M & S medium with 9 g per litre of agar. Replace IAA and kinetin with 2 mg per litre of 6-benzylaminopurine (BAP) and 0.01 mg per litre of 1-naphthylacetic acid (NAA). (BAP and NAA are available from Aldrich.)

© Cambridge University Press 1994

Table 2.2 The composition of M & S medium, amounts and functions

Substance (amount/mg l^{-1})	Examples of functions
ammonium nitrate (1650)	nitrogen source
calcium chloride (440)	enzyme cofactor
magnesium sulphate (370)	enzyme cofactor
potassium dihydrogenphosphate (170)	buffer
potassium nitrate (1900)	nitrogen source
boric acid (6.2)	involved in cell division
cobalt chloride (0.02)	involved in cell division
copper sulphate (0.025)	component of cytochromes
manganese sulphate (22.3)	enzyme cofactor
potassium iodide (0.83)	enzyme cofactor
sodium molybdate (0.25)	component of nitrogenase
zinc sulphate (8.6)	enzyme cofactor
disodium ethylenediamine tetra-acetate (EDTA) (37.3)	no specific function
iron sulphate (28.8)	enzyme cofactor
glycine (2.0)	protein synthesis
meso-inositol (100)	for building cell walls
nicotinic acid (0.5)	component of NAD/NADP
pyridoxine hydrochloride (0.5)	enzyme cofactor
thiamine hydrochloride (0.1)	enzyme cofactor
sucrose (30 000)	carbon and energy source
indol-3-yl acetic acid (IAA) (2.0)	growth hormone
kinetin (0.2)	growth hormone

Answers to questions

1. Callus (undifferentiated tissue) should be observable after about a week. Development after this depends upon the plant, and the ratio of the different types of growth hormones.
2. Other organisms can grow in this culture medium, competing with and contaminating the plant tissue culture.
3. Students would not be expected to know the function of every item on the list (completed in Table 2.2), unless it were to be used as part of a longer assignment on plant nutrition. Some components have several functions, e.g. the potassium dihydrogenphosphate acts as a buffer and a phosphate source.
4. Horticultural or rare plants, e.g. orchids, whose unique characteristics could be lost by breeding. Crops prone to disease – cassava, rice or wheat (disease-resistant strains can be cloned and grown under carefully monitored conditions before being planted out).
5. 2,4-D is toxic to broad-leaved (dicot) weeds at amounts that are not toxic to grasses and cereals (monocots). The hormone leads to uncontrolled growth of the weeds, which leads to their eventual destruction.

© Cambridge University Press 1994

Suggestions for project work

1. Other plants could be grown. Carrot, potato and tobacco are reported to give good results, or students could try out their own choices.
2. Investigate the effect of changing the level of plant hormones on root and shoot growth. In general, a high auxin/cytokinin ratio leads to rooty growth and vice versa.
3. Plan and cost out a larger-scale tissue culture operation to provide plants of a chosen type for a market garden, school fair or demonstration plot.

Assessment guide

Experimental skills

1. Careful selection and excision of plant tissue (2).
2. Sterilisation of plant tissue (2).
3. Sterile transfer of plant material to medium (2).
4. Organisation of apparatus on bench to maintain safety and sterility (2).
5. Maintenance of sterile conditions through the duration of the experiment; looking after plants over period of observation (2).

© Cambridge University Press 1994

Experiment 3
Tryptic digest of casein

Aim
This experiment shows you the first steps in one approach to protein sequence analysis.

Introduction
The primary structure of a protein can be determined by splitting the polypeptide chain into smaller peptide fragments, whose sequence can then be determined by further chemical techniques. Enzymes that cut polypeptide chains are known as proteases. A protein will give a different set of peptide fragments depending upon which protease is used. Here we use trypsin, which cuts a polypeptide chain only at the carboxy side of a lysine or arginine residue. This obviously limits the extent to which the chain is degraded. This is useful for obtaining information about the primary structure. Complete degradation of the polypeptide only provides information about amino acid composition, and says nothing about the sequence.

The fragments produced from the tryptic digest must be separated from each other before they can be sequenced. This is usually done by a combination of electrophoresis and chromatography, which provides a 'fingerprint' like the one shown in Figure 3.1. A more modern method would be to use high-performance liquid chromatography (HPLC) as shown in Figure 3.2.

In this experiment you use one-dimensional paper chromatography to separate the peptide fragments produced by the tryptic digestion of the milk protein, casein. (See also Chapter 2, *Biochemistry for advanced biology*.)

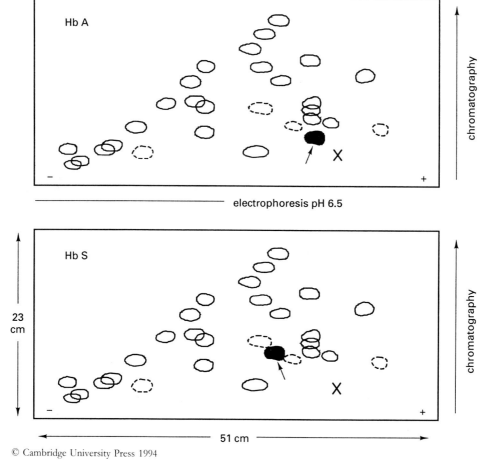

Figure 3.1 Fingerprints of human haemoglobins A and S. The black spot marked with an arrow indicates the peptide that differs in the two fingerprints.

© Cambridge University Press 1994

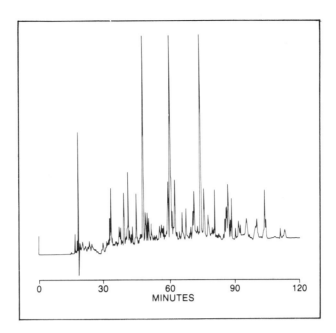

Figure 3.2 Tryptic digest of casein. HPLC analysis.

Equipment and materials

- two small, stoppered glass bottles of volume around $2\,cm^3$
- two $1\,cm^3$ graduated pipettes
- water bath at $37\,°C$
- $500\,cm^3$ measuring cylinder
- an oven at $110\,°C$
- capillary tubes
- filter paper
- aluminium foil
- sellotape
- trypsin solution (0.1 per cent)
- casein solution (1 per cent)
- solvent mixture of butan-1-ol, ethanoic acid, water (4/1/5 volume ratio)
- buffer at pH 7.0
- ninhydrin spray

SAFETY

Trypsin may cause irritation to the skin and eyes. Avoid direct contact or inhalation, and clean up spillages immediately.

Ninhydrin should be used in a fume cupboard.

Experimental procedure

1. Add $0.1\,cm^3$ of trypsin solution to $1\,cm^3$ of the casein solution in one stoppered bottle. In the other, put $1\,cm^3$ casein and $0.1\,cm^3$ buffer as a control.
2. Incubate the bottles in the water bath for 30 minutes.
3. Meanwhile, cut a 20 cm strip of filter paper to fit the measuring cylinder and place solvent in the cylinder to a depth of about 1 cm,

covering with a foil lid. Mark a base line in *pencil* 1 cm from the end of the strip of filter paper.

4. Note the appearances of the bottles in the water bath over the incubation period.
5. After 30 minutes, remove the bottles from the water bath and make small, concentrated spots of the tryptic digest mixture and the control on the base line. It is important that the spots should be as concentrated as possible. The way to do this is to 'build up' a spot by applying a spot no more than 5 mm in diameter, allowing it to dry and then applying another spot over the first. For best results, this should be repeated at least four times, allowing the spot to dry before the next is applied each time.
6. Attach the strip of paper to the foil lid with sellotape, as shown in Figure 3.3, and carefully lower it into the measuring cylinder, taking care that the solvent does not reach the base line.
7. The solvent will now travel up the filter paper, carrying the peptides from the tryptic digest with it. This may take several hours.

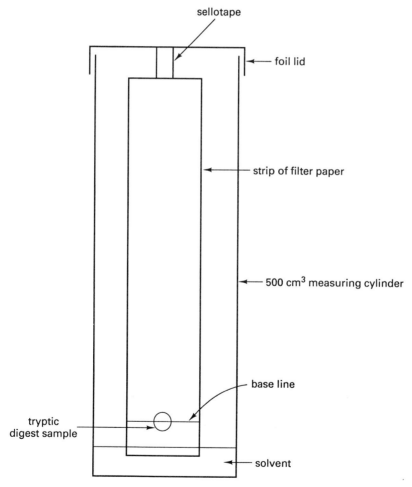

Figure 3.3 Chromatography of tryptic digest.

8. When the solvent has reached about 1 cm from the top of the paper, remove the paper from the measuring cylinder and allow it to dry thoroughly in a fume cupboard.
9. When it is dry, spray with the ninhydrin solution and place in the oven for 5–10 minutes.

© Cambridge University Press 1994

QUESTIONS

1. What did you observe as your tryptic digest progressed? How many peptides would you expect if casein has about 280 amino acids, assuming that the frequency of occurrence of a particular amino acid is one in 20?
2. How many peptides can you detect in your chromatogram, and how many are detectable in the HPLC profile in Figure 3.2. Account for any differences.
3. Find the peptide that differs in the two haemoglobin samples in Figure 3.1.
4. This peptide was shown to contain eight amino acids, which differ by only one residue in the two types of haemoglobin. What experiments could you carry out to show which amino acid differs?
5. How could you produce shorter peptides from the casein sample?
6. How does your control sample compare with that which has undergone tryptic digestion?
7. How would the following affect your results – (a) increasing the time of incubation to several hours, (b) increasing the temperature of the water bath to 50 °C?

Teacher/technician notes

Equipment and materials

- two small, stoppered glass bottles of volume around $2\,cm^3$
- $1\,cm^3$ graduated pipettes
- water bath at $37\,°C$
- $500\,cm^3$ measuring cylinder
- an oven at $110\,°C$
- capillary tubes (drawn out to a fine capillary in a bunsen burner for applying spots to chromatogram)
- trypsin solution (0.1 per cent in buffer)
- casein solution (1 per cent low-fat Marvel or other dried milk powder in water)
- solvent mixture of butan-1-ol, ethanoic acid, water in 4/1/5 volume ratio)
- buffer at pH 7.0 (0.067 M ($9.1\,g\,l^{-1}$) potassium dihydrogenphosphate and 0.067 M disodium hydrogenphosphate ($9.5\,g\,l^{-1}$) in a 2/3 volume ratio)
- ninhydrin spray in propanone (0.2 per cent solution or Philip Harris spray pack)
- filter paper (Whatman No. 1)
- aluminium foil
- sellotape

Experimental procedure

The chromatograms take several hours to run – an overnight run may be more convenient. Generally, the longer the chromatogram the better the separation, although the peptides are likely to run as a long streak rather than as discrete spots under these conditions. Ninhydrin gives purple colours with peptides.

Answers to questions

1. The tryptic digest clarifies after a few minutes while the control stays milky. If Lys and Arg occur at a frequency of one in 10 (two out of 20 amino acids), then we would expect, statistically, 28 out of the 280 amino acids of casein to be Lys or Arg, which would give 29 peptides if all possible cleavages occurred.
2. One-dimensional chromatography gives a relatively poor separation, but demonstrates that peptides have been produced in comparison with the control. Optimum separation requires two-dimensional chromatography, fingerprinting (electrophoresis followed by chromatography) or HPLC.
3. Refer to Figure 3.2 – the peptides that differ are to be found in the bottom right-hand corner.
4. Amino acid analysis (hydrolyse with 6 M HCl for 24 hours) and separation with two-dimensional chromatography. Compare with known amino acid samples or use ion-exchange chromatography with standard amino acids. (There is valine instead of glutamic acid in the abnormal haemoglobin.)
5. Use a mixture of proteolytic enzymes to get further cleavage.

© Cambridge University Press 1994

6. Control gives no colour with ninhydrin as it has not been cleaved (protein does not generally react with ninhydrin).
7. (a) More proteolysis with longer reaction time. (b) Trypsin degrades at 50 °C.

Assessment guide

Experimental skills

1. Digestion experiments set up with careful attention to enzyme/casein ratios and maintenance of water bath temperature and timing (2).
2. Preparation of set-up for chromatography including correct positioning of sample spot (2).
3. Build-up of small concentrated spot (2).
4. Ninhydrin spray used in fume cupboard, paper well sprayed and incubated for correct length of time. Production of chromatogram with visible evidence of peptides alongside control (2).
5. Experiment carried out in logical and well organised way (2).

© Cambridge University Press 1994

Experiment 4
Extracting DNA from onions

Aim
This experiment gives you experience of extracting biologically important materials from tissue. It also demonstrates the physical nature of DNA.

Introduction
The extraction of DNA from cells is the first step of many experiments in molecular biology. Treatment with detergent is an effective way of breaking open the cells and their nuclei to release the contents. The soluble cellular extract contains proteins and nucleic acids, including DNA. Sodium chloride helps the nucleic acids to coalesce, while an enzyme breaks down the protein. Finally, cold ethanol precipitates out the crude DNA.

DNA is rather a fragile molecule. It is only 2 nm wide, but even the smallest DNA molecules, those from viruses, have a length about 1000 times greater than this. Like a glass fibre, DNA breaks easily. Even stirring with a glass rod will fragment it into shorter pieces. The hydrogen bonds holding its two strands together in a double helix are easily broken by heating. The two strands separate abruptly at a certain temperature – usually between 60 and 80 °C – in a process known as **denaturation** or **melting**, which can also be brought about by acid or alkali treatment.

(See also Chapter 3, *Biochemistry for advanced biology*.)

Equipment and materials
- liquidiser
- vegetable knife and chopping board
- water bath at 60 °C
- ice bath
- thermometer
- 100 cm^3 measuring cylinder
- two 500 cm^3 beakers
- filter funnel
- filter paper
- stop clock
- glass rod
- boiling tube and rack, or 100 cm^3 beaker
- dropping pipette
- onion
- sodium chloride/detergent solution
- protease solution
- ice-cold ethanol
- 4 per cent sodium chloride
- pH paper

Experimental procedure
1. To extract the onion tissue, first chop the onion into small pieces and cover with 100 cm^3 sodium chloride/detergent solution in a beaker.

© Cambridge University Press 1994

2. Stir the mixture and incubate in the water bath for 15 minutes.
3. Cool in the ice bath for 5 minutes.
4. Liquidise the mixture for 5 seconds at high speed.
5. Filter the mixture into the second beaker.
6. To extract the DNA, first add five drops of protease to the filtered tissue extract in a boiling tube or a small beaker and mix well.
7. Now carefully pour a layer of ice-cold ethanol on *top* of the mixture. Leave to stand for a few minutes.
8. The precipitated DNA should form a web of tangled fibres, which can be drawn up from the tube by carefully rotating a glass rod at the interface between the two layers (see Figure 4.1).

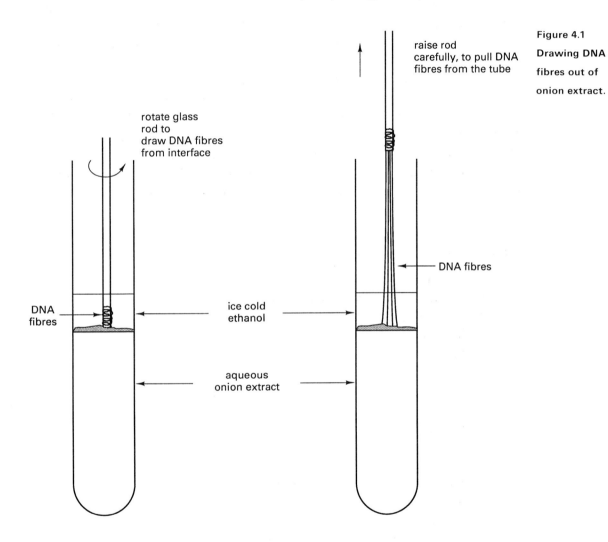

Figure 4.1 Drawing DNA fibres out of onion extract.

QUESTIONS

1. What is the main contaminant of DNA prepared in this way? How could it be removed?
2. What is the function of the protease?
3. Why must the time in the liquidiser be minimised?
4. What precautions would you take to protect the DNA from degradation in further experiments with it?
5. What results would you get if you tested DNA with universal indicator? (Try this; resuspend the DNA in 4 per cent sodium chloride first.)
6. What does the physical appearance of the DNA prepared in this way tell you about its molecular nature?

© Cambridge University Press 1994

Teacher/technician notes

Equipment and materials

- liquidiser
- vegetable knife and chopping board
- water bath at 60 °C
- ice bath
- thermometer
- 100 cm^3 measuring cylinder
- two 500 cm^3 beakers
- filter funnel
- filter paper (coffee filter paper is best)
- stop clock
- glass rod
- boiling tube and rack, or 100 cm^3 beaker
- dropping pipette
- onion
- sodium chloride/detergent solution (use 30 g sodium chloride, 100 cm^3 any cheap washing-up liquid dissolved in 1 litre distilled water)
- protease solution (use Novo Neutrase^{TM} from NCBE)
- ice-cold ethanol (95 per cent, chill in freezer in a plastic bottle overnight)
- 4 per cent sodium chloride
- pH paper

The filtrate can be stored in a fridge for up to 2 days.

Answers to questions

1. RNA. It could be removed by an enzyme, ribonuclease, which degrades it into soluble fragments.
2. It degrades the proteins associated with the DNA (histones) into soluble fragments by cleaving the peptide bonds.
3. DNA is fragmented if the extract is in the liquidiser for too long.
4. Work in buffer to avoid extremes of pH, work at low temperature, chilling reagents on ice as you work, avoid long delays.
5. DNA is acidic, so a red colour is seen.
6. The fibrous nature of the DNA isolated suggests that it is a polymer.

Assessment guide

Experimental skills

1. Careful preparation of onion extract – shown by attention to detail of instructions, e.g. timing of liquidiser step (2).
2. Clear filtrate obtained (2).
3. Careful addition of ethanol with formation of two layers (2).
4. Works quickly and methodically (2).
5. Demonstrates production of DNA fibres (2).

© Cambridge University Press 1994

Experiment 5
Food testing

Aim
After this investigation you should be able to identify and estimate amounts of protein, reducing sugars and fat in a given food sample. You should appreciate how the relative contributions of these nutrients to a given food are worked out.

Introduction
Manufacturers often display nutritional information on food packets. This is not required by law, but if it is included it must be presented as follows: the values are given in grams contained in 100 grams of the product. Usually the following are listed – energy value, in both kilocalories and kilojoules, total carbohydrate, sugars, protein, fat, dietary fibre and sodium.

    The food tests done in the school laboratory are the basis of some of the analyses done in industry, although in the latter more sophisticated equipment is used, in the interests of accuracy. Reducing sugars are analysed either by titration with Fehling's solution or by high-performance liquid chromatography. Protein content is determined by calculation from the amount of nitrogen in the food, found by chemical analysis. Fat is extracted into petroleum ether and weighed after evaporation of the solvent. The bright yellow colour that sodium compounds give in a flame is used as a basis for working out the sodium content of foods. Finally, moisture content is calculated by heating a weighed sample of the food at 500 °C for several hours and measuring the loss in mass, while dietary fibre content is given by subtracting the reducing sugar content from the total carbohydrate content.

    (See also Chapters 4 and 5, *Biochemistry for advanced biology*.)

Equipment and materials
- oven set at 100 °C
- test tubes with rack
- dropping pipette
- balance
- bunsen burner
- test tube holder
- 25 cm^3 conical flask
- small spatula
- colorimeter with filter and tubes
- nickel crucible
- 1 cm^3 graduated pipettes
- 100 cm^3 volumetric flask
- dilute hydrochloric acid
- sodium hydrogencarbonate
- pH paper
- selection of foods – with packets and food labels if possible
- Clinitest tablets or Clinistix strips
- petroleum ether 60–80
- Bradford protein assay solution

© Cambridge University Press 1994

SAFETY

Petroleum ether is flammable. Keep it well away from naked flames, and work in a fume cupboard if possible.

Clinitest tablets contain solid sodium hydroxide, which is corrosive, so avoid skin contact and handle with forceps only.

Bradford protein assay solution is acidic, so avoid skin contact.

Experimental procedure

Determination of fat

1. Add 5 cm^3 petroleum ether to an aqueous suspension of a weighed sample (about 1 g) of the food and shake in a stoppered test tube.
2. Remove the top petroleum ether layer, which will have extracted most of the fat, with the dropping pipette and transfer to a weighed conical flask.
3. Repeat this extraction.
4. Allow the petroleum ether to evaporate in the fume cupboard. (It takes about an hour, so do this test first if you want to complete it in one session; otherwise leave it overnight.) Then reweigh the flask.
5. Calculate the percentage of fat in the food sample.

Determination of moisture content

1. Place a sample of food in a nickel crucible and weigh; 2–3 g should be enough. Put into the oven and leave overnight.
2. Cool and reweigh. Calculate the percentage moisture in the sample.

Determination of reducing sugars

1. Suspend a weighed (about 0.1 g) sample in 1 cm^3 water.
2. Add one Clinitest tablet or dip a Clinistix strip into the solution. Use the charts provided with the tablets or strips to read off the percentage of reducing sugar in the sample.
3. If your sample is too concentrated, dilute it (making a note of how much you dilute it) until you get an 'on-scale' reading. Calculate the percentage of reducing sugar in your sample, remembering to take account of any dilution that you have made.
4. If the test was negative, you should now test for sucrose. Make up another sample of the food, as above, and boil with 0.5 cm^3 dilute hydrochloric acid for one minute, cool and add small amounts of solid sodium hydrogencarbonate until there is no more effervescence to neutralise (check with pH paper). This breaks down any sucrose to fructose and glucose. Repeat the test, as above.

Determination of protein

1. Make a suspension of a weighed (about 0.1 g) sample of food in 10 cm^3 water.
2. Dilute this by 100 by dispensing 1 cm^3 into the 100 cm^3 volumetric

© Cambridge University Press 1994

flask with the graduated pipette and making up to the mark with distilled water.

3. Make up a sample in a colorimeter tube as follows: $0.2\,cm^3$ Bradford assay reagent and $0.8\,cm^3$ of your diluted protein sample.
4. Read and record the optical density at 595 nm, using the appropriate filter, blanking the colorimeter first on a control made of $0.2\,cm^3$ Bradford assay reagent and $0.8\,cm^3$ water.
5. Read off the protein concentration from the graph provided (Figure 5.1). Multiply this value by 5/4 to account for the dilution made in the colorimeter tube and by 1000 to account for the 100-fold dilution and for the fact that you took only $1\,cm^3$ of the $10\,cm^3$ of your protein extract. This gives you the amount of protein in your food sample. Note that, if you are working with a high-protein food, you may need to dilute by 1000 instead of 100 to obtain a reading on scale.

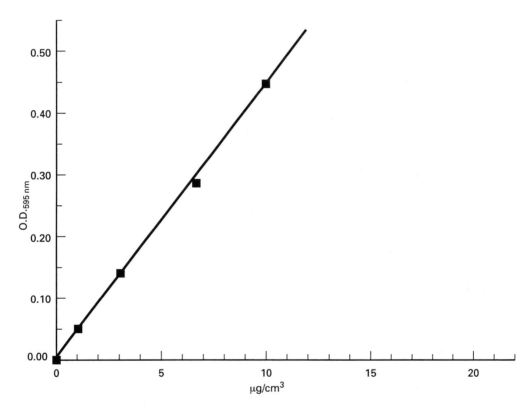

Figure 5.1 Typical standard curve for the Bradford protein assay.

QUESTIONS

1. Calculate the percentage composition of the food(s) you have investigated and present your results as a pie chart.
2. Write a brief description of the food(s) with reference to nutritional qualities.
3. How do your results compare with the data on the label provided with the food?
4. Identify sources of inaccuracy in the tests you have done and make suggestions for improvement.
5. Do your values add up to 100 g? What else do you think is in your food sample?

© Cambridge University Press 1994

Teacher/technician notes

Equipment and materials

- oven set at 100 °C
- test tubes with rack
- dropping pipette
- balance
- colorimeter with filter and tubes
- nickel crucible
- bunsen burner
- test tube holder
- 25 cm^3 conical flask
- small spatula
- pH paper
- 1 cm^3 graduated pipettes (ideally four per student or group)
- 100 cm^3 volumetric flask
- dilute hydrochloric acid (1 M)
- sodium hydrogencarbonate
- selection of foods – with packets and food labels if possible (good examples include soft drinks with added sugar and with added sweetener, margarine with its high fat content, dried milk powder, which has a high protein content, and eggs)
- Clinitest tablets or Clinistix strips (Philip Harris)
- petroleum ether 60–80
- Bradford protein assay solution from: Solution A: 100 cm^3 95 per cent ethanol, 200 cm^3 88 per cent phosphoric acid, 350 mg Brilliant Blue G (from Aldrich). Solution B: 425 cm^3 distilled water, 15 cm^3 95 per cent ethanol, 30 cm^3 88 per cent phosphoric acid. To make up the assay solution, add 30 cm^3 of solution A to the quantity of solution B specified above, filter through Whatman No. 1 filter paper and store in the fridge.

Answers to questions

1-3. Answers depend on foods used.
4. Moisture content may be too low if sample is not heated to constant mass. Larger sample gives more accurate result, especially if water content is low. Clinitest tablets and Clinistix strips are also semi-quantitative, depending on assessment of colour by eye. Fat may not all be extracted into petroleum ether; multiple extraction could improve accuracy. In protein assay, the calibration curve should ideally be redetermined for each assay and will depend upon the instrument used.
5. Insoluble (dietary) fibre and minerals probably account for the rest of the food sample.

Assessment guide

Data handling
1. Convert raw data into percentage composition of foods tested (4).
2. Presentation of percentage composition as a pie chart (2).
3. Translation of percentage composition into conclusion about nutritional quality of the foods (2).
4. Able to criticise methodology (2).

© Cambridge University Press 1994

Experiment 6
The chromatography of plant pigments

Aim
To appreciate the use of paper chromatography as a technique for the separation and identification of a mixture of plant pigments.

Introduction
Chromatography – the word means 'writing with colour' – is one of the most important techniques in biochemistry for separation and purification of the chemicals (such as pigments and proteins) that are found in cells.

In chromatography, a mixture of chemicals is applied to an absorbent material, which is then placed in contact with a solvent. As the solvent soaks through the material, it carries the components of the mixture along with it at rates that differ according to their solubility and how strongly they are absorbed. This process leads to separation of the components from each other.

In paper chromatography, the absorbent is the cellulose of the paper. Because pigments are coloured substances, their separation on cellulose can be observed by eye.

(See also Chapters 2 and 6, *Biochemistry for advanced biology*.)

Equipment and materials

- scissors
- mortar and pestle
- spatula
- filter paper
- filter funnel
- test tubes and rack
- fine capillary tubes
- large boiling tubes with rubber bungs
- drawing pins
- fine sand
- plant material – use as wide a range as possible, including fruits and flowers if they are available
- propanone
- acidified methanol
- solvent A (a mixture of petroleum ether and propanone)
- solvent B (a mixture of glacial ethanoic acid, hydrochloric acid and water)

SAFETY

Organic solvents are flammable and their vapours may be harmful.

Acidified methanol and *solvent B* are corrosive – wear eye protection and gloves when dispensing them.

© Cambridge University Press 1994

Experimental procedure

1. Extract photosynthetic pigments by grinding finely chopped leaves or grass covered with propanone in a mortar and pestle. The addition of a spatula of fine clean sand helps to extract the maximum amount of pigment. Filter the extract into a small labelled test tube.
2. Extract flower and fruit pigments as above, but use acidified methanol instead of propanone.
3. Cut rectangular strips of chromatography paper that can be suspended freely in the boiling tubes as shown in Figure 6.1a, drawing a fine *pencil* line 1 cm from one end.
4. Dip the fine capillary into the pigment extract and transfer it to a spot mid-way along the base line. Allow it to dry.
5. Repeat this process three or four times, allowing the spot to dry each time between each application. Best results are obtained from small concentrated spots of extract (the spot containing the photosynthetic pigments should be dark green if you are to see all the pigments).
6. Place solvent in the boiling tube(s) to a depth of about 2 cm, using A for photosynthetic pigments and B for fruit and flower pigments.
7. Pin the strip of chromatographic paper to the bung, as shown, and lower it carefully into the tube so that the end dips into the solvent. The solvent level must lie *below* the spot.
8. Allow the chromatogram to run until the solvent front lies just below the bung – this takes 1–2 hours.
9. Carefully remove the chromatogram, marking the level reached by the solvent, and allow it to dry.

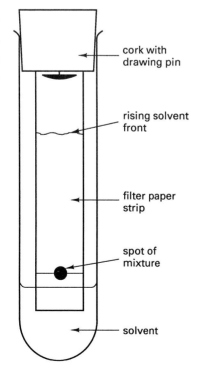

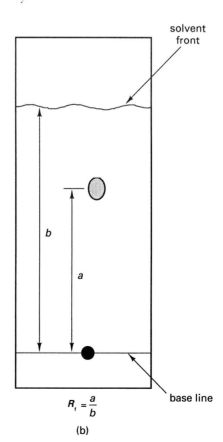

$$R_f = \frac{a}{b}$$

Figure 6.1 (a) Chromatography set-up. (b) Calculation of R_f value.

© Cambridge University Press 1994

THE CHROMATOGRAPHY OF PLANT PIGMENTS

QUESTIONS

1. Figure 6.1b illustrates how the R_f value for each component on the chromatogram is calculated. Note that R_f is the ratio of the distance moved by the component to the distance moved by the solvent. An R_f of 0.5, for example, means that the component has moved half of the distance moved by the solvent. The R_f value gives a measure of the rate at which the component has moved and a way of identifying it. How many components can you distinguish on your chromatograms and what are their colours? Calculate the R_f value for each one and identify them from the data in Table 6.1. How could you verify your results.
2. What factors might cause your R_f values to vary from those given above for the same pigment?
3. How could you refine the technique to improve poor separation?
4. How would you detect invisible spots on a chromatogram?
5. How could you use this technique on an industrial scale to isolate the red dye shikonin from the roots of the wild herb *Lithospermum erythrorhizon*?
6. Sketch the chromatogram you would expect from a chestnut leaf (a) in summer and (b) in autumn, and account for any differences.

Table 6.1 R_f values of pigments

Type of pigment	R_f value
Photosynthetic pigments (in solvent A)	
Chlorophyll *b*	0.45
Chlorophyll *a*	0.65
Xanthophylls	0.70
Phaeophytin	0.80
Carotene	0.95
Flower and fruit pigments (in solvent B)	
Pelargonidin	0.70
Malvidin	0.60
Cyanidin	0.50
Petunidin	0.45
Delphinidin	0.30

© Cambridge University Press 1994

Teacher/technician notes

Equipment and materials

- scissors
- mortar and pestle
- spatula
- filter paper (Whatman No. 1)
- filter funnel
- test tubes and rack
- fine capillary tubes (pull out in a hot bunsen flame to make a fine pipette for application of spots onto paper)
- large boiling tubes with rubber bungs
- plant material (see below)
- drawing pins
- fine sand
- propanone
- acidified methanol (1 M hydrochloric acid and methanol in a 1:9 ratio)
- solvent A (a mixture of petroleum ether and propanone in a 9:1 ratio)
- solvent B (a mixture of glacial ethanoic acid, hydrochloric acid and water in a 30:3:10 ratio)

Plant material Use as wide a range as possible, including fruits and flowers if they are available. Spinach works particularly well. Leaves could include some that are not green, e.g. copper beech, or are variegated, such as geranium or *Coleus*. Other materials may include broad bean and other seed coats, flowers such as geranium, cornflower, chrysanthemum and roses, and grapes and strawberries.

Answers to questions

1. Students should appreciate that the R_f values are a way of comparing spots on different-sized chromatograms. All the photosynthetic pigments listed in Table 6.1 are visible under these conditions. The xanthophylls run together. Strawberries and geraniums both contain pelargonidin, roses and red cabbage contain cyanidin. To verify their assignment of a spot, they would run it alongside a pure sample of the substance as a reference, under the same conditions. Point out that two pigments coincidentally could have the same R_f value and could be distinguished only by doing some other form of analysis or perhaps running the chromatogram under different conditions.
2. R_f values depend upon temperature, the type of absorbent (different grades of paper, for example) and solvent composition. Length of paper or time taken to run the chromatogram do not affect R_f values.
3. Two-dimensional chromatography, or different solvent mixtures, can be discussed. A longer strip of paper would also give better separation between spots with similar R_f values.
4. Look at under a UV lamp, or spray with a reagent that gives a coloured product.
5. Use a large sheet of paper and a solvent tank. Apply root extract as

a concentrated band, cut out and elute with a suitable solvent. For large-scale work, use a column of cellulose.
6. In summer all the photosynthetic pigments are present; in autumn chlorophyll and phaeophytin break down, leaving xanthophylls and carotene.

Assessment guide

Data handling

1. Correct calculation of R_f values (2).
2. Identification of pigments from Table 6.1 (2).
3. Appreciation of need for verification of identity and suggestions as to how this can be done (2).
4. Appreciation of factors that affect R_f values (2).
5. Understanding of limitation of the experiment, sources of error and suggestions for improvement (2).

© Cambridge University Press 1994

Experiment 7
The effect of sunlight on photosynthesis

Aim
This experiment gives you experience in designing an experiment to discover the effect of altering variables on a biological process. It also demonstrates the effect of light on the rate of photosynthesis.

Introduction
In **photosynthesis**, the energy of sunlight drives a complex chain of chemical reactions that manufacture glucose from carbon dioxide. The rate of production of the by-product, oxygen, can be used to find out how quickly a plant is photosynthesising.

Plants can adapt their photosynthetic apparatus to climatic conditions. C_3 **plants**, found in temperate climates, fix carbon at lower temperatures and higher levels of carbon dioxide than do C_4 **plants**; the latter have another set of chemical reactions that allow more efficient use of carbon dioxide. C_4 plants include tropical plants such as sugar cane, which keep their stomata closed to prevent water loss at high temperature. Carbon dioxide levels within the plant are low and C_4 photosynthesis takes over to allow for this.

Understanding how photosynthetic activity may change in a warmer world with more carbon dioxide in the atmosphere may be crucial to our long-term survival. Some scientists have suggested that higher temperatures and increased carbon dioxide levels may favour C_3 plants. Others have pointed out that higher rates of photosynthesis can lead to higher-yielding crops. Under laboratory conditions, many major food crops, including wheat, cassava, potatoes and rice, show increases in yield between 10 and 50 per cent when the amount of carbon dioxide in the air is increased.

(See also Chapter 6, *Biochemistry for advanced biology*.)

Equipment and materials

- apparatus for measuring rate of photosynthesis, as shown in Figure 7.1
- lamp
- stop clock
- filter pump
- flask with side-arm and cork
- thermometer
- metre rule
- *Elodea* (Canadian pondweed) or mint
- sodium hydrogencarbonate

Experimental procedure

1. Produce a written plan of how to investigate the effect of light intensity on photosynthesis. You should consider the following points:
   (a) How to measure the rate of oxygen evolution. Possibilities include counting gas bubbles, trapping oxygen in a test tube and measuring the volume, drawing oxygen into a syringe. You

© Cambridge University Press 1994

could measure the volume evolved in a given time, or the time taken to evolve a given volume.

(b) Selection of apparatus. It can take several hours to produce a full test tube of gas. The apparatus in Figure 7.1a measures smaller volumes. Or you could use a large amount of plant material as in Figure 7.1c to increase the volume of oxygen.

(c) How to ensure that the gas collected is oxygen and not air bubbling out of the water.

(d) How to supply an adequate amount of carbon dioxide to speed up photosynthesis.

(e) How to keep other factors, such as temperature, constant.

(f) How to get enough results to arrive at a conclusion in the time available.

2. Check your plan with your teacher and carry out the experiment.

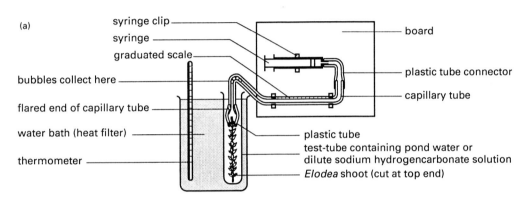

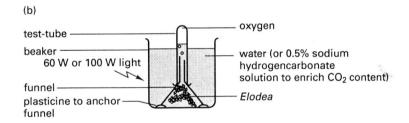

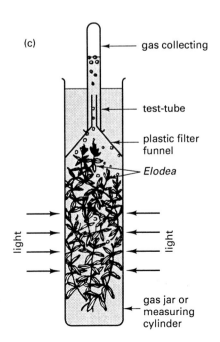

Figure 7.1 Choice of apparatus for measuring the effect of light intensity on photosynthesis.

Assessment guide

Experimental design
1. Appropriate selection of apparatus and plant material (2).
2. Logical and detailed written plan (2).
3. Appreciation of need to keep temperature and water composition constant and details of how to do this (2).
4. Method for measuring rate of oxygen evolution given (2).
5. Reasonable estimation given of time needed and plans for how to get data required by either teamwork or timetable of when readings will be taken (2).

Experiment 8
Aerobic and anaerobic respiration in yeast

Aim
After carrying out this experiment, you should appreciate that yeast can respire both in the presence and in the absence of oxygen, giving different products in each case.

Introduction
Some organisms need oxygen to extract energy from food during respiration while others do not – in fact, oxygen may even be toxic to them. **Aerobic respiration** and **anaerobic respiration** follow different chemical pathways after **glycolysis** and their end-products differ. Yeast, and many other organisms, can respire both aerobically and anaerobically. Although maximum energy is extracted from glucose and other carbon sources by aerobic fermentation, useful commercial products are obtainable from anaerobic fermentations. These include ethanol – used in alcoholic drinks, but also as a fuel – and solvents such as acetone and butanol, which are used in the chemical industry.

(See also Chapter 6, *Biochemistry for advanced biology*.)

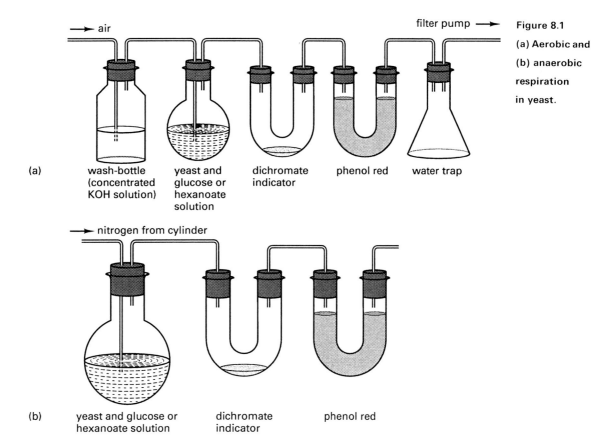

Figure 8.1 (a) Aerobic and (b) anaerobic respiration in yeast.

Equipment and materials
- U-tubes, water trap and connectors set up as in Figures 8.1a and b
- two 500 cm^3 round-bottomed flasks
- four 250 cm^3 glass beakers
- spare wash bottle
- yeast

© Cambridge University Press 1994

38 EXPERIMENT 8

- concentrated potassium hydroxide
- filter pump
- dichromate indicator
- glucose
- phenol red indicator
- phosphate buffer, pH 7
- source of nitrogen gas
- sodium hexanoate

SAFETY

> *Potassium hydroxide* is corrosive. Wear gloves and safety goggles when handling it.
>
> *Sodium hexanoate* may contain traces of hexanoic acid. Wear gloves to weigh it out.

Experimental procedure

1. Weigh out 1 g glucose and dissolve it in 40 cm^3 phosphate buffer in a glass beaker. Weigh out 8 g yeast and suspend it in buffer in another beaker. Transfer the yeast to the round-bottomed flask and bubble air through the apparatus for 10 minutes (do not attach the rest of the apparatus yet).
2. Assemble the rest of the apparatus as shown in Figure 8.1a, placing 5 g dichromate indicator in one U-tube, phenol red indicator to a height of 3 cm on each side of the other U-tube. Place the potassium hydroxide in the wash bottle, making sure that it covers the inlet tube by at least 2 cm.
3. Add glucose to the round-bottomed flask and connect to the rest of the apparatus. If the yeast foams too much, attach the spare wash bottle between the round-bottomed flask and the U-tube to contain any overflow.
4. Adjust the rate at which air is bubbled through the apparatus with the tap of the filter pump. Too much bubbling causes the yeast to foam, too little will stop air reaching all the yeast cells, so that some will respire anaerobically.
5. Observe any changes in the indicator flasks for the next few hours.
6. For respiration under anaerobic conditions, repeat with the apparatus shown in Figure 8.1b, but first bubble nitrogen through the yeast suspension for several minutes.
7. Both experiments can be repeated replacing glucose with sodium hexanoate as a substrate – this is the salt of a fatty acid. Use 0.75 g of sodium hexanoate.

QUESTIONS

1. Why is the potassium hydroxide solution used?
2. What does each indicator show?
3. Explain the changes you observed in the indicator flasks and say what the products of respiration are in each case.
4. Suggest some other substrates that could be used as a substitute for glucose.

© Cambridge University Press 1994

Teacher/technician notes

Equipment and materials

- U-tubes, water trap and connectors set up as in Figures 8.1a and b
- two 500 cm^3 round-bottomed flasks
- four 250 cm^3 glass beakers
- spare wash bottle
- yeast (any dried bakers' yeast works well)
- concentrated potassium hydroxide (4 M solution, 224 g per litre)
- filter pump
- dichromate indicator (see below)
- glucose
- phenol red indicator (this turns from red to yellow when carbon dioxide is produced)
- phosphate buffer, pH 7 (see Experiment 3 for recipe)
- source of nitrogen gas (use a nitrogen gas cylinder (British Oxygen) secured safely to the bench)
- sodium hexanoate (see below)

Dichromate indicator Dissolve 0.5 g potassium dichromate indicator in 25 cm^3 concentrated sulphuric acid – *CARE*, this is a corrosive mixture – and add this to 25 cm^3 distilled water. Add this to chromatography-grade aluminium oxide with stirring until it is just moist and very pale yellow.

Sodium hexanoate Hexanoic acid (from Aldrich) is insoluble, so should be converted into the more soluble sodium salt as follows: add 44.6 g of the acid to 100 cm^3 4 M sodium hydroxide and stir. Evaporate this solution to obtain the salt.

Because these experiments take several hours, the class should divide into groups and share their results.

Answers to questions

1. The potassium hydroxide absorbs any carbon dioxide from the air, so we can be sure that any detected by the phenol red indicator came from the respiring yeast.
2. Potassium dichromate turns from yellow to green when ethanol is produced. Phenol red turns from red to yellow when carbon dioxide is produced.
3. Aerobic respiration of glucose produces carbon dioxide from about 15 minutes onwards, but only trace amounts of ethanol if aeration is efficient. Anaerobic respiration produces ethanol and some carbon dioxide, so both indicators are positive.

    Sodium hexanoate is the salt of a six-carbon fatty acid, while glucose is a six-carbon carbohydrate. This experiment compares fatty acid and carbohydrate metabolism. Since fatty acids can form acetyl CoA, they can enter the TCA cycle and form carbon dioxide, and this will be observed under aerobic conditions; but since acetyl CoA cannot be converted into pyruvate, no carbon dioxide or ethanol can be formed under anaerobic conditions.

© Cambridge University Press 1994

4. Sucrose, starch and other carbohydrates could be used. Cheap sources such as molasses and corn syrup are often used in industry (pre-treatment with enzymes may be necessary).

Assessment guide

Data handling

1. Accurate record of indicator changes from all experiments (2).
2. Links observations to type of respiration (2).
3. Links observations to type of substrate (2).
4. Demonstrates understanding of respiration in overall conclusion (2).
5. Understands use of control and can generalise results to other substrates (2).

Experiment 9
Questionnaire on the Human Genome Project

Aim
This investigation will help you to review your knowledge of human genetics. You will also practise communicating scientific ideas by devising an information sheet and a questionnaire. Analysis of the results of the questionnaire will require you to consider ways of presenting data accurately and clearly.

Introduction
The **Human Genome Project** is an international scientific effort that aims to construct a map of our chromosomes. This will show the location of all our genes. The project will also work out the base sequence of the entire genome.

The consequences for human health could be immense. First, we could discover genes that are linked to the development of common diseases such as cancer, when they mutate. Locating these genes could lead to tests for tendencies to inherit such diseases. Secondly, so much of the human genome is unknown territory that we are bound to uncover thousands of new genes. Some of these could control human characteristics such as height. Others may turn out to be involved in qualities that are not currently thought of as having a genetic component, such as musical or mathematical ability. Finally, we could learn more about the biochemical basis of diseases that are not very well understood at the moment, such as **Huntington's chorea** and **motor neurone disease**. This could lead to more effective treatment – in particular, to **gene therapy**.

While detailed knowledge of our genetic make-up could enable us to live happier, fuller lives, some people may prefer not to be burdened with this kind of information. But there is likely to be pressure from employers and insurance companies for people to discover their genetic status. At the Edinburgh Science Festival in 1991, social scientist Hilary Rose warned that society is not prepared for the consequences of the Human Genome Project. 'We must make sure that people's feelings are taken as seriously as their genes', she said.

(See also Chapter 7, *Biochemistry for advanced biology*.)

Equipment
- Background reading material: 'In the beginning was the genome', Roger Lewin, *New Scientist*, 21 July 1990; 'Britain and the human genome', John Galloway, *New Scientist*, 28 July 1990; 'Human blueprint: is it better not to know?', Andy Coghlan, *New Scientist*, 2 May 1992
- access to computer (optional)
- graph paper
- calculator

© Cambridge University Press 1994

Procedure

Design a questionnaire 'pack' that includes the following.

1. A quiz, testing your subjects' existing knowledge of human genetics. Answers to the quiz.
2. A hand-out of no more than one side of A4 explaining clearly what the Human Genome Project is, and how it will affect your subjects.
3. A glossary including the following terms: gene, genome, DNA, recessive, dominant, chromosome, carrier, autosomal, X-linked.
4. The questionnaire itself, with questions designed to elicit your subjects' response to the prospect of the Human Genome Project and the effect that it will have on their lives.

Some ideas to get you started

1. A couple with two children suffering from sickle cell anaemia (SCA) interviewed recently on the radio admitted that they had misunderstood when they were told that their children would have a one-in-four chance of being born with the disease. When their first child was born with SCA, they assumed that the next three would be born without it. How would you find out how well your subjects understand this issue?
2. In a recent survey of 1000 people in the UK, in connection with screening for cystic fibrosis (CF) carrier status, 76 per cent had heard of the disease, 35 per cent knew it was inherited, 18 per cent knew that carriers need not have a family history of the disease, 39 per cent said they would consider not having children if they were carriers, and 26 per cent thought they would terminate an affected pregnancy.
3. You should research the material for this project, starting with the background reading.
4. Think carefully about the subject group you will approach and gear your material to their age and background. You could try the following – your tutor group, your teachers, people in the street, students in a different school or college, a local community group.
5. Make the quiz straightforward – perhaps multiple choice – so that you can analyse the results easily. In the questionnaire, have some questions that require a simple yes/no/don't know (like an opinion poll), again for simple analysis. Don't ask loaded ('Do you agree that the Human Genome Project will lead to mass murder of innocent foetuses?') or vague ('Is the Human Genome Project a good thing?') questions. Be specific ('Would you want to know if you were a CF carrier?'). Ask some questions that need more thoughtful replies, too, so that you gain more insight into people's views.
6. Think about the best way to present your data. You could use a pie chart, table or bar chart. You should also write a clear summary. You could present your work on the school/college noticeboard or in a letter to a newspaper or magazine.

© Cambridge University Press 1994

Teacher notes

This exercise is suitable for small group work – each individual can take responsibility for one part of the questionnaire package, with input from the others in the group.

Obtain any necessary permissions before students interview subjects. This is a good opportunity to have students build up confidence by communicating outside of their peer groups – make use of any contacts you have outside the school/college.

Assessment guide

Experimental design
1. Content of questionnaire package (2).
2. Presentation of questionnaire package (2).
3. Selection of subjects and collection of sufficient data (2).
4. Analysis and presentation of data – percentages, tables, pie charts, using computer if available (2).
5. Clear detailed summary with evidence of background reading (2).

© Cambridge University Press 1994

Experiment 10
Looking at DNA fingerprints

Aim
In this experiment you will learn how DNA fingerprints are analysed to obtain useful information about an individual's genetic status. You will also discover some of the uses of DNA probes, as well as revising and perhaps extending your knowledge of genetic diseases.

Introduction
The hydrogen bonding between base pairs in DNA has allowed the development of some powerful diagnostic techniques. Base-pairing makes it possible to pick out one gene from a background of nearly 3000 million base pairs of human DNA.

Although there are now many different types of DNA analysis, the following steps are common to most of them, and have been used in the examples on which you will be working. This procedure is known as **Southern blotting** (Figure 10.1).

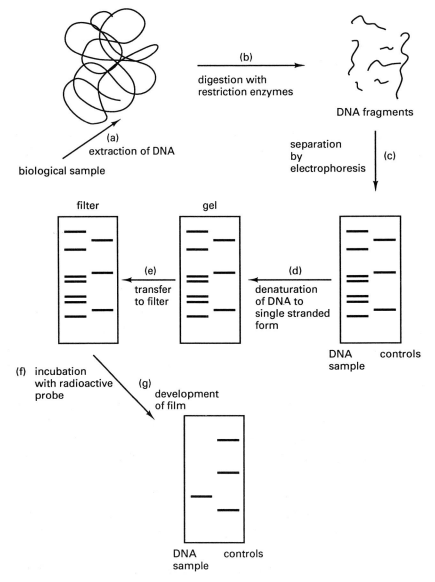

Figure 10.1 Southern blotting. (Note: until step (g) the bands are invisible – they have been marked in here to illustrate the principles involved.)

© Cambridge University Press 1994

(a) DNA is extracted from a biological sample such as blood or saliva.
(b) The DNA is chopped up into smaller fragments – usually a few kilobases long (a kilobase, or kb, is 1000 bases) – with restriction enzymes. These cut the DNA at a particular sequence. For example, the enzyme TaqI recognises and cuts only at the sequence TCGA.
(c) The fragments are separated from each other by gel electrophoresis. A sample is applied to a tray of gel. An electric current is then passed through the gel. The fragments move along with the current, with the smaller ones moving faster than the larger ones.
(d) The gel is treated with alkali to denature the DNA fragments.
(e) The single-stranded DNA fragments are transferred to a filter. This is done by placing the filter – which is made of either nylon or a form of cellulose – on top of the wet gel and placing weights on top of this arrangement for several hours to allow the DNA fragments to soak through to the filter.
(f) The filter is incubated with a radioactively labelled probe. The probe is a length of DNA that searches out its complementary sequence and binds to it.
(g) The filter is placed in contact with a photographic film. Radio-labelled bands – the ones which have bound to the probe – show up, while others remain invisible.

(See also Chapters 7 and 8, *Biochemistry for advanced biology*.)

In Figure 10.1, the probe is part of the human Factor IX gene (Factor IX is a protein involved in blood clotting). The analysis shows up the Factor IX gene in the DNA sample.

Procedure
Study each of the following examples, and answer the questions linked to each one.

Example 1. Detecting carriers of haemophilia B, an X-linked disease

Haemophilia B results from mutations in the Factor IX gene, which is found on the X chromosome. Heterozygous females have one mutant and one normal Factor IX allele. They are carriers, but do not have the disease themselves. Males who inherit the mutant allele suffer from haemophilia – a disorder in which the blood does not clot properly.

The analysis depends upon differences – called **polymorphisms** – along the DNA sequence of a pair of chromosomes. Sometimes these will abolish or create a sequence that is cut by a restriction enzyme. For example, the enzyme TaqI recognises the sequence TCGA and cuts DNA at this position. Imagine two alleles, identical in their sequence except for one base, and suppose in this region the sequence of one is TCGA, while that of the other is TCCA. This single base difference – G instead of C – is a polymorphism. TaqI will cut the first allele, but not the second. So the first allele would have two short DNA fragments where the second would have a single, longer one.

Polymorphisms that can affect the length of the fragments produced in a restriction enzyme digest in this way are called **restriction fragment length polymorphisms** (RFLPs). Figure 10.2 shows what happens when TaqI reacts with a

46 EXPERIMENT 10

sequence containing an RFLP (as described above). The DNA fragments are picked up by a probe whose sequence spans that of the RFLP.

Figure 10.3a shows the results of DNA analysis of a normal family, using the probe shown in Figure 10.2.

QUESTIONS
1. Is the female heterozygous or homozygous for this allele?
2. Why do all the samples show a 5.3 kb fragment?

RFLPs are often inherited along with a disease if they are close to the defective gene. Here the RFLP is actually within the Factor IX gene, but in a non-coding region (so it can also occur in a normal family). Figure 10.3b shows the DNA analysis of a family where the son suffers from haemophilia, using the same RFLP and probe.

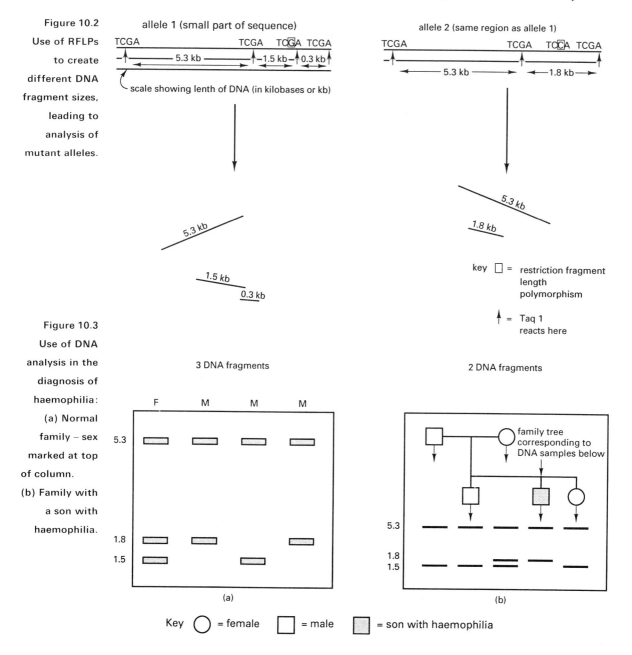

Figure 10.2 Use of RFLPs to create different DNA fragment sizes, leading to analysis of mutant alleles.

Figure 10.3 Use of DNA analysis in the diagnosis of haemophilia: (a) Normal family – sex marked at top of column. (b) Family with a son with haemophilia.

QUESTIONS

3. Is the mother here heterozygous or homozygous?
4. Which DNA fragment is specifically associated with haemophilia?
5. Is the daughter a carrier?

Example 2. Detecting carriers of cystic fibrosis, an autosomal recessive disease

One in 20 Europeans carries a mutant gene that leads to cystic fibrosis. CF leads to chronic lung disease, and sufferers need constant treatment. Many die before or in their mid-twenties. Figure 10.4 shows a DNA analysis based on an RFLP that gives a 950 base fragment on one allele and 600 and 350 base fragments on the other.

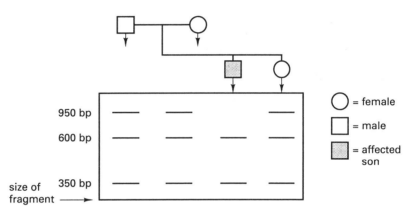

Figure 10.4 DNA analysis of a family with cystic fibrosis.

QUESTIONS

6. Are the parents homozygous or heterozygous for this pair of alleles?
7. What is the chance of any of their children having cystic fibrosis?
8. Is the daughter a carrier?
9. What are her chances of producing a child with CF if her partner is (a) a carrier, (b) not a carrier and (c) suffers from CF? Draw family-tree diagrams showing genotypes to illustrate your answer.

A new test has been developed which can detect the four commonest CF mutations in the British population. In the ARMS test, the patient's DNA sample is split between two test tubes. Each tube contains four different DNA probes. If the 'normal' probes are numbered 1–4, and the 'mutant probes 1*–4*, then the first tube contains probes 1, 4, 2* and 3* while the second contains probes 2, 3, 1* and 4*. Figure 10.5 shows results from patients A–F.

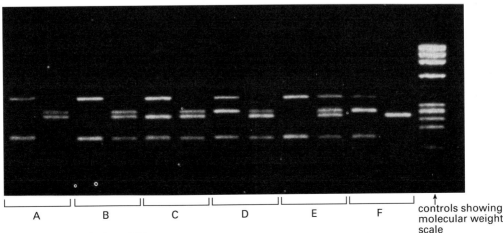

Figure 10.5 Results from an ARMS test.

© Cambridge University Press 1994

QUESTIONS
10. A is normal. Bands for sequences 1–4 run from the top to the bottom of the diagram. A heterozygote has the normal pattern, plus one band in duplicate. Which patient is a heterozygote, and for which mutation?.
11. Some patients have two mutations. Which are they and what are the mutations?
12. Which patient is a homozygote and why?
13. Could A still be a carrier of CF?

Example 3. Huntington's chorea, an autosomal dominant disease

Huntington's chorea (HC) is a brain disease that leads to progressive mental deterioration. Unlike other inherited diseases, it is not apparent at birth. People suffering from it only notice the symptoms in middle age – usually after they have had their families.

The defective gene in HC has recently been discovered. Several RFLPs that are inherited with the disease have been identified and form the basis of a test for the defective gene.

QUESTIONS
14. A woman in her late twenties learns that her father has HC. She and her partner are planning a family. Draw diagrams to show the risk of an affected foetus if (a) she has the HC gene and (b) she does not have the HC gene. What assumptions do you make? (Note: if she has the gene, she will, in time, develop the disease herself – as it is dominant – which may also affect her decision to have a family.)
15. The region of chromosome 4 that includes the RFLP and the HC gene is shown in Figure 10.6. Draw the patterns that you would see on Southern blotting analysis using a DNA probe that spans the RFLP if the woman (a) has the defective HC allele and (b) has a normal allele. What assumptions do you make?

Figure 10.6 Use of linked RFLP to detect Huntington's chorea. (Note: actual data would be far more complex – this has been simplified to demonstrate the principles involved more easily.)

Example 4. Establishing identity

DNA fingerprinting, which is carried out to establish identity, uses many probes. The result shows a complex pattern of bands, which looks rather like the bar code on the items you buy in the supermarket. The technique assumes that the chances of any individual fingerprint being identical are very low. So a match between samples is said to have a high chance of establishing identity. Figure 10.7a shows DNA fingerprints of a rape victim, of the semen specimen and of three suspects.

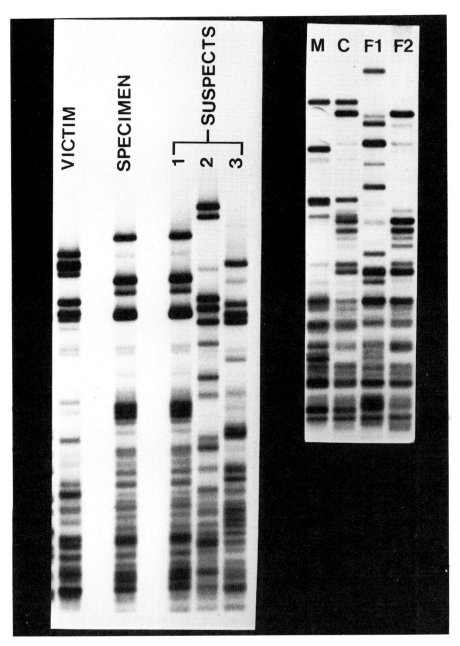

Figure 10.7 DNA fingerprints: (a) from a rape victim, the semen specimen and three suspects; (b) from a family where paternity was disputed, M = mother, C = child and F1 and F2 are the potential fathers.

QUESTION

16. Which suspect matches the specimen?

Figure 10.7b shows fingerprints from a family where there was a dispute about who was the father of the child (C). The child inherits DNA from the mother (M) and the father (F1 or F2).

QUESTIONS

17. Who is the father of this child?
18. What other uses can you think of for this technique?
19. Why do you think DNA fingerprinting has been criticised recently by the legal profession?

© Cambridge University Press 1994

Teacher notes

Answers to questions

1. She is heterozygous because she has bands that come from both alleles (5.3, 1.8 and 1.5 kb).
2. They all have the 5.3 kb band because there is no RFLP within this fragment.
3. The mother is again heterozygous, with all three bands, as in question 1.
4. The 1.8 kb band is associated with haemophilia – the son has inherited this from his mother. The mutant allele has one less restriction enzyme site than the normal.
5. She is not a carrier because she has not inherited the 1.8 kb band.
6. The parents are heterozygotes who have the allele that gives the 950 base fragment and the allele with the 600 and 350 base fragments.
7. The chance of an affected child are one in four (mutant alleles must be inherited from both parents).
8. Yes, she is a heterozygote, like her parents.
9. Chance is 1/4 if partner also a carrier, zero if not, and 1/2 if he is affected.
10. B is heterozygous for 4.
11. C (3 and 4), D (2 and 4), E (1 and 4).
12. F.
13. Yes, A may have a rare mutation not covered by these probes.
14. Risk is 1/2 if she is affected, zero if not. This assumes that her mother and partner do not carry the defective HC gene.
15. See Figure 10.8. Assume that this RFLP is always inherited with the defective HC gene – in fact, the probability depends on the distance between the RFLP and the gene on the chromosome.

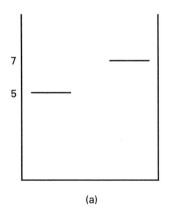 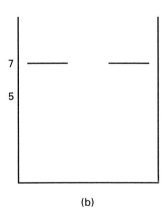

Figure 10.8 Solution to question 15: (a) carries the HC gene (associated with 5.3 kb fragment) but (b) does not carry the HC gene. (Note: the subject is a heterozygote, with a normal and a mutant allele; homozygotes for the HC gene are statistically very unlikely.)

© Cambridge University Press 1994

16. Suspect 1.

17. The father is F2.

18. Other criminals, if they have left biological specimens at the scene of crime; identification of corpses; bone marrow transplants have been successful if fingerprints belong to donor rather than patient.

19. Matches occur more often than has been suggested, as the human population does not mate at random; samples may not have been kept in optimum conditions and may be contaminated or degraded; careless laboratory technique, as shown by lack of reproducibility between laboratories.

Assessment guide

Data handling

1. Understanding of method (questions 2, 4 and 15) (2).
2. Interpreting DNA fingerprints (questions 1, 5, 6, 7, 8 and 9) (2).
3. Drawing valid conclusions (questions 10, 11, 12, 13, 17 and 19) (2).
4. Presenting data (questions 9 and 14) (2).
5. Appreciation of scope and limitation of method (questions 14, 15, 18 and 19) (2).

© Cambridge University Press 1994

Experiment 11
The extraction of polyphenol oxidase from mushrooms

Aim
To appreciate the delicate nature of biological materials, acquire skill in isolating one component from a complex system, and evaluate the extent of purification achieved.

Introduction
The isolation of biologically important molecules, such as enzymes, from cells is an important part of practical biochemistry. First, the cells must be broken open, by chemicals such as detergents that degrade cell membranes. This releases the cell contents, usually into a buffer solution. Centrifugation at different speeds can remove organelles such as nuclei and mitochondria – these remain in the solid residue, or **pellet**, while free proteins, including enzymes, remain in the top liquid layer, or **supernatant**.

Proteins are precipitated out from this liquid by the addition of substances such as propanone or polyethylene glycol. They can then be redissolved in a smaller volume of buffer. This crude extract should show enzyme activity if a substrate is added and the mixture incubated at a suitable temperature.

Further purification of proteins obtained by these techniques can be the most difficult part of the operation, but is necessary if new enzymes are to be characterised and investigated. Often, chromatography or **electrophoresis** is involved. The latter technique uses the fact that proteins of different sizes move at different rates in an electric field (chromatography is discussed in Experiment 6). If **antibodies** to the required enzyme are available, these can be fixed onto a column of resin and the crude extract passed down it. The required protein should latch onto its antibody, leaving the impurities to flow through the column. The protein is later removed from the column – usually by passing through a solution of low pH, which breaks the bonds between the protein and its antibody. This technique is called **affinity chromatography**.

In this experiment, **polyphenol oxidase** is extracted from mushroom tissue. This enzyme is widely distributed in fruits and vegetables. It causes browning when their tissue is exposed to air, as it oxidises compounds containing a phenol group to dark-coloured compounds called polyphenols. The dark colour of tea comes from polyphenols.

(See also Chapters 2 and 9, *Biochemistry for advanced biology*.)

Equipment and materials
- scalpel
- mortar and pestle
- muslin square – about $12 \times 12 \, cm^2$
- three $100 \, cm^3$ beakers
- one $500 \, cm^3$ volumetric flask
- two $10 \, cm^3$ measuring cylinders
- two $25 \, cm^3$ measuring cylinders
- one $500 \, cm^3$ measuring cylinder
- one $1 \, cm^3$ graduated pipette
- three $2 \, cm^3$ pipettes

- two small crucibles
- stop clock
- rack and test tubes
- ice bath
- balance
- centrifuge and tubes
- colorimeter with blue filter
- tubes to fit colorimeter
- water bath at 30 °C
- oven at 110 °C
- 10 g mushrooms
- 2 g fine white sand
- phosphate buffer, pH 7
- catechol solution
- propanone

SAFETY

Propanone is inflammable and should be used in the fume cupboard.

Catechol is corrosive – use gloves when handling it.

Experimental procedure

Work as quickly as possible to avoid denaturation of the enzyme.

1. Chop mushroom tissue and grind with sand and 5 cm³ buffer, gradually adding another 5 cm³ buffer.
2. Strain through muslin into 100 cm³ beaker, centrifuge for 2 minutes.
3. Decant the supernatant into a measuring cylinder and make up to 20 cm³ with buffer.
4. Transfer to another 100 cm³ beaker and cool to 5 °C in ice bath.
5. Add the same volume of ice-cold propanone, dropwise, stirring and keeping cool.
6. Centrifuge this suspension for 2 minutes and resuspend the pellet in 10 cm³ of buffer.
7. Centrifuge for 1 minute. Make the supernatant up to 10 cm³ with distilled water. This is the crude enzyme extract.
8. For assay, this must be diluted. Pipette 1 cm³ of the crude extract into a 500 cm³ volumetric flask and make up to the mark with distilled water.
9. Finally, weigh a small sliver of mushroom – about 0.05 g. You are going to use this to see how much polyphenol oxidase activity there is in a mushroom before the enzyme is extracted.
10. Make up four test tubes as follows.
    (A) Control for calibrating the colorimeter: 2 cm³ buffer and 4 cm³ distilled water.
    (B) Control for the assay: 2 cm³ buffer, 2 cm³ 0.02 M catechol and 2 cm³ water (no enzyme).
    (C) 2 cm³ buffer, 2 cm³ 0.02 M catechol and 2 cm³ water with the mushroom sliver.

© Cambridge University Press 1994

54   EXPERIMENT 11

(D) 2 cm³ buffer, 2 cm³ 0.02 M catechol and 2 cm³ crude enzyme extract (add this last, just before measuring).

11. First zero the colorimeter with tube A.
12. Incubate tubes B, C and D for 5 minutes at 30 °C, then immediately measure their absorbances.
13. Finally, to compare the activity of enzyme in tubes C and D, the dry masses must be found. First weigh out about 2 g mushroom tissue into a crucible and heat overnight in the oven to drive off the water. The value you obtain from this result (see below) is the percentage dry mass of a typical mushroom (like many vegetables, mushrooms contain a lot of water) and it can be used in the calculations that follow.

    Also take 5 cm³ of the crude extract and heat it in the same way. This will give you the dry mass of the enzyme extract.

    Allow both crucibles to cool and then weigh them.
14. Calculate the specific activity of the polyphenol oxidase in fresh mushroom tissue and in the crude extract from the following formula:

$$\text{specific activity in units per gram} = \frac{(\text{absorbance of sample}) - (\text{absorbance of tube B})}{\text{dry mass of sample}}$$

To obtain the dry mass of the mushroom sliver, first calculate the percentage dry mass of mushroom tissue, A. This is given (using data from step 13) by:

$$A = \frac{\text{dry mass of mushroom}}{\text{fresh mass of mushroom}} \times 100$$

Then the dry mass of the mushroom sliver – which you need to work out the activity of the enzyme – is given by:

$$\frac{A \times \text{fresh mass of mushroom sliver}}{100}$$

The dry mass of tube D is given by:

$$\frac{(\text{dry mass of crude extract from step 13}) \times 2}{5}$$

(The dry mass from step 13 is multiplied by 2 and divided by 5 to account for the fact that 2 cm³ are used in the assay and 5 cm³ are used in step 13.) Multiply the answer for the crude extract by 500 to account for the dilution.

QUESTIONS

1. How does the polyphenol oxidase activity of fresh mushrooms compare with that of the crude enzyme extract?
2. What makes the enzyme particularly prone to denaturation in this experiment, and what steps do you take to avoid this?
3. Why did you not dry the mushroom sliver itself rather than the separate 2 g of mushroom tissue?
4. Which fruits and vegetables do not contain polyphenol oxidase (i.e. do not go brown)?
5. How could you protect fruits and vegetables from browning?
6. What steps would you take to purify this protein further?

© Cambridge University Press 1994

Teacher/technician notes

Equipment and materials

- scalpel
- mortar and pestle
- muslin square – about $12 \times 12\,cm^2$
- three $100\,cm^3$ beakers
- one $500\,cm^3$ volumetric flask
- two $10\,cm^3$ measuring cylinders
- two $25\,cm^3$ measuring cylinders
- one $500\,cm^3$ measuring cylinder
- one $1\,cm^3$ graduated pipette
- three $2\,cm^3$ pipettes
- two small crucibles
- stop clock
- rack and boiling tubes
- ice bath
- balance (weighing to 3 dp)
- centrifuge and tubes
- colorimeter with blue filter
- tubes to fit colorimeter
- water bath at $30\,°C$
- oven at $110\,°C$
- 10 g mushrooms (these should be as fresh as possible)
- 2 g fine white sand
- phosphate buffer, pH 7 ($0.067\,M$ ($9.1\,g/l^{-1}$) potassium dihydrogenphosphate and $0.067\,M$ ($9.5\,g/l^{-1}$) disodium hydrogenphosphate mixed in a 2:3 ratio; allow $30\,cm^3$ per student or group)
- catechol solution ($0.02\,M$ ($2.2\,g/l^{-1}$); catechol is obtainable from Aldrich)
- propanone

The amount of time that the enzyme solution spends at room temperature should be minimised – to prevent denaturation. It can be stored in a fridge if delays are anticipated.

Answers to questions

1. Sample results:
   absorption of control B, 0.02
   absorption of tube C, 0.16; corrected value 0.14
   absorption of tube D, 0.16; corrected value 0.14
   mass of fresh mushroom, 2.110 g; mass when dried, 0.28 g
   percentage dry weight of tissue is 13.2
   dry mass of $5\,cm^3$ crude extract, 0.213 g
   mass of sliver is 0.023 g
   dry mass of tube C is $0.132 \times 0.023 = 0.003\,g$
   specific activity is $0.14/0.003 = 46.67$ units per gram of tissue
   dry mass of tube D is $0.213 \times 0.4 = 0.085\,g$

© Cambridge University Press 1994

specific activity of crude extract is $(0.14/0.085) \times 500 = 823.5$ units per gram of tissue; this shows that some concentration has occurred.
2. The presence of propanone – an organic solvent – makes the enzyme prone to denaturing. Working quickly and using buffers and low temperatures all help to minimise denaturation.
3. The error in the weighing would be more significant at small masses.
4. Melons and tomatoes.
5. Boil them, keep in acid – to denature polyphenol oxidase.
6. Electrophoresis, ion-exchange chromatography, or affinity chromatography if an antibody is available.

Further work

1. Measure amounts of polyphenol oxidase in different fruits and vegetables.
2. Attempt further purification by ion-exchange chromatography.

Assessment guide

Experimental skills

1. Follows procedures correctly, with little or no guidance required (2).
2. Works in systematic and organised way (2).
3. Can use balance and colorimeter correctly to take the required readings (2).
4. Shows appreciation of delicate nature of enzymes by working quickly and using buffers, ice baths and refrigeration where required (2).
5. Shows due regard for safety (2).

© Cambridge University Press 1994

Experiment 12
The action of pepsin on egg white

Aim
To demonstrate the action of a protease (pepsin) and some of the characteristics of enzyme action.

Introduction
Proteases are enzymes that hydrolyse proteins into smaller peptide fragments. They work best on denatured proteins. When eggs are cooked, the protein is rapidly denatured by heat. Egg whites contain several proteins – the most abundant being ovalbumin. There are also traces of lipid.

When pepsin works on egg-white suspension – the physical state in which it would be in the stomach, where pepsin is one of several digestive enzymes – it rapidly clears or clarifies it. This is because the peptides formed by enzyme action are water-soluble, while the denatured proteins are not. The peptide bonds in proteins can also be broken by boiling the protein in concentrated acid for several hours.

(See also Chapters 2 and 9, *Biochemistry for advanced biology*.)

Equipment and materials

- stop clock
- bunsen burner, tripod and gauze
- 500 cm^3 beaker
- test tubes and rack
- boiling tube and holder
- water bath at 25 °C
- 1 cm^3 syringe
- 2 cm^3 syringe
- 100 cm^3 conical flask
- 10 cm^3 measuring cylinder
- dropping pipette
- distilled water
- pepsin (40 mg sample in tightly stoppered bottle)
- egg
- 2 M hydrochloric acid

SAFETY

Hydrochloric acid is corrosive. Use safety glasses when handling it and deal with spills immediately by neutralising it with sodium hydrogencarbonate before wiping up.

Experimental procedure
1. Break the egg and separate its white into a small beaker.
2. Put 25 cm^3 distilled water into a boiling tube and add four drops of egg white, stirring the suspension thoroughly after each drop.
3. Heat the test tube over a medium bunsen flame until the suspension turns milky. The proteins are now denatured.
4. Cool the tube under a running cold water tap.

5. Make an acidified egg-white suspension by adding $0.5\,cm^3$ 2 M HCl to the $10\,cm^3$ measuring cylinder using the $1\,cm^3$ syringe. Then add egg-white suspension to the $10\,cm^3$ mark.
6. Add $20\,cm^3$ distilled water to the bottle containing pepsin.
7. Transfer a $1\,cm^3$ portion of this to a test tube and denature by heating it in a boiling water bath (use the $500\,cm^3$ beaker) for 5 minutes. Allow to cool.
8. Label four tubes A, B, C and D. Add $2\,cm^3$ egg-white suspension to A, B and C. Add $3\,cm^3$ distilled water to D and $1\,cm^3$ distilled water to C. Incubate the tubes in the water bath for 5 minutes. Make a copy of Table 12.1 to record your results.

Table 12.1

A	B	C	D
$2\,cm^3$ egg white	$2\,cm^3$ egg white	$2\,cm^3$ egg white	$3\,cm^3$ water
$1\,cm^3$ pepsin	$1\,cm^3$ denatured pepsin	$1\,cm^3$ water	

9. Now add $1\,cm^3$ pepsin solution to A and the $1\,cm^3$ portion of denatured pepsin to B and start the stop clock.
10. Note and record the appearance of the tubes at suitable intervals for the next hour.
11. Finally add another $2\,cm^3$ egg-white suspension to tube A. Record the time taken to clarify.

QUESTIONS
1. Comment on the use of controls in this experiment.
2. Which properties of enzymes are demonstrated by this experiment?
3. Account for any faint cloudiness remaining after clarification.
4. Would you expect pepsin to clarify a suspension of (a) olive oil in water, (b) cooked meat particles and (c) milk?
5. How would you degrade the protein in egg white if no enzymes were available?

Teacher/technician notes

Equipment and materials

- stop clock
- bunsen burner, tripod and gauze
- $500\,cm^3$ beaker
- test tubes and rack
- boiling tube and holder
- water bath at $25\,°C$
- $1\,cm^3$ syringe
- $2\,cm^3$ syringe
- $100\,cm^3$ conical flask
- $10\,cm^3$ measuring cylinder
- dropping pipette
- distilled water
- pepsin (40 mg sample in tightly stoppered bottle; this quantity clarifies the egg white within 15 minutes)
- egg (one egg will provide enough protein for several students)
- 2 M hydrochloric acid

The egg-white suspension and the pepsin solution will keep for a few days in the fridge, but should then be discarded.

Answers to questions

1. The control C contains no enzyme, suggesting that it is the enzyme that produces clarification. The water control D provides a clear blank against which to measure clarification.
2. Enzymes are catalysts and are not used up in the reaction, as shown by the clarification of a second portion of protein added to tube A. They are denatured by heating – no clarification is seen in tube B, which contains denatured enzyme.
3. The slight cloudiness is due to lipids, which are present in egg white and are not acted on by proteases and are not soluble in water (this shows enzyme specificity).
4. Pepsin acts on meat and milk suspensions (protein and fat) but not olive oil (lipid).
5. Heat with acid for several hours.

Assessment guide
Observing and recording

1. Makes observations in a systematic way (2).
2. Sufficient data (at least five observations of each tube) recorded (2).
3. Data recorded in an acceptable format, e.g. a table (2).
4. Details of appearance of tubes (cloudiness, floating particles, residual lipid opalescence) recorded (2).
5. Checks temperature and records temperature of water bath (2).

© Cambridge University Press 1994

Experiment 13
Investigating the thermal stability of chymosin

Aim
This experiment is an opportunity for you to plan an investigation into the effect of temperature on enzyme activity.

Introduction
Many enzymes denature at temperatures above 40 °C. So, although the rates of all chemical reactions increase with temperature, the rates of enzyme reactions have an upper limit. However, some organisms, such as the bacteria that live in hot springs, have enzymes that can withstand temperatures up to 100 °C. One of these enzymes is **Taq polymerase**, which comes from the species *Thermophilus aquaticus*. Taq polymerase catalyses the multiplication of DNA molecules in the **polymerase chain reaction** (PCR). One stage of PCR takes place at about 70 °C. Thanks to Taq polymerase, there is no need to keep adding fresh enzyme to the reaction mixture, and PCR has now been fully automated. This makes it easy to amplify tiny DNA samples for forensic work and prenatal diagnosis of genetic disease.

Heat-stable enzymes have other uses, such as in detergent mixtures. So it is important to establish the temperature profile of any new enzyme that is extracted. This is done by measuring the enzyme's activity over a range of temperature, as you will do in this investigation.

(See also Chapter 9, *Biochemistry for advanced biology*.)

Equipment and material
- water bath, which can be set to a range of temperatures
- thermometer (0–100 °C)
- test tubes and rack
- stop clock
- 1 cm^3 graduated pipette
- 10 cm^3 measuring cylinder
- chymosin
- milk
- phosphate buffer, pH 7.0

Planning your investigation
In this experiment, the substrate is soluble casein in milk. A methionine–phenylalanine peptide bond in this protein is broken by the protease enzyme chymosin, leading to the formation of an insoluble form of casein.

First, investigate the action of chymosin on milk to determine (a) suitable enzyme and substrate concentrations to use (you do not need to dilute the milk, but will need to dilute the enzyme) and (b) how to assay the action of the enzyme.

After these preliminary experiments, make a written plan of how you will investigate the temperature profile of chymosin. Include your preliminary observations and details of any control experiments. Check your plan with your teacher and carry out the experiment.

© Cambridge University Press 1994

QUESTIONS

1. Plot your results on a graph, to show how the activity of chymosin varies with temperature. Explain the shape of your graph.
2. Estimate the temperature at which chymosin becomes denatured.
3. What was the purpose of your control experiment?
4. How would you speed up the coagulation of milk other than by altering the temperature?
5. Give an industrial use for chymosin. What characteristics should an industrial chymosin have?

Teacher/technician notes

Equipment and materials

- water bath, which can be set to a range of temperatures
- thermometer (0–100 °C)
- test tubes and rack
- stop clock
- 1 cm^3 graduated pipette
- 10 cm^3 measuring cylinder
- chymosin (see below)
- milk (any kind; about 5 cm^3 with 1 cm^3 1/100 chymosin will coagulate milk at 30–40 °C in 5 to 10 minutes)
- phosphate buffer (made up as described in Experiment 3 and used to dilute the chymosin)

Chymosin

Pure chymosin (Maxiren^{TM}) is available from the NCBE. This is a genetically engineered chymosin, used in the manufacture of vegetarian cheese. It should be stored at 4 °C and diluted 1/100 in distilled water just before use. Students could be provided with 1 cm^3 aliquots of the concentrated solution to investigate the dilution needed themselves. You could also use commercial rennin (rennet), which contains pepsin and chymosin.

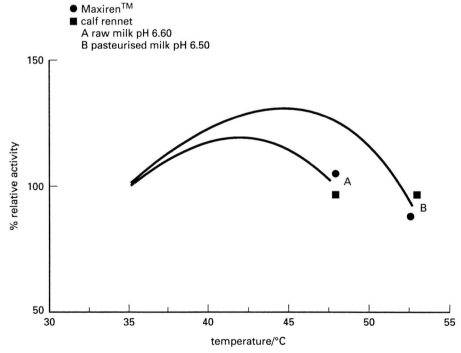

Figure 13.1 Effect of temperature on chymosin.

Answers to questions

1. See Figure 13.1.
2. Chymosin denatures at 55 °C.
3. There should be a control without enzyme to show that coagulation is a result of enzyme action.
4. Increase the enzyme concentration, work at pH optimum.
5. Cheese making. Should be a pure product, if used for the food industry, and a wide pH and temperature range are desirable as well as high specificity so cheese does not get broken down into small peptides.

Assessment guide

Experimental design

1. Preliminary investigation of assay of activity (e.g. which point of coagulation process to take as end-point) and concentration of enzyme (2).
2. Detailed and logical plan of work (2).
3. Wide range (say, 15–60 °C) and sufficient readings (at least five) to give temperature profile (2).
4. Inclusion of control experiment (2).
5. Explanation of how results will be turned into conclusions – for example, by plotting reciprocal of coagulation time against temperature of the water bath (2).

© Cambridge University Press 1994

Experiment 14
The effect of pH on enzyme activity

Aim
After carrying out this experiment, you should appreciate the sensitivity of enzymes to extremes of pH.

Introduction
The shape of an enzyme molecule is crucial to its functioning. This shape depends upon chemical bonds holding the molecule together in its tertiary structure. Some of these are ion pairs, formed between basic and acidic side chains in amino acid residues. Residues at the active site may also be affected by pH.

The pH profile of an enzyme is a useful guide to its effectiveness in various situations. For instance, pepsin and carboxypeptidase are both digestive proteases, but have quite different pH profiles. Pepsin, with its optimum pH of 2.0, is active in the acidic conditions of the stomach, while carboxypeptidase, which is most active at pH 7.0, is found in the small intestine. Pepsin has two glutamic acid residues at the active site, one of which must be in the acid form.

Most fungal amylases, such a those used in bread making, have a pH maximum between 4.5 and 5.0. This makes then unsuitable for inclusion in washing powders. Termamyl™, which is extracted from *Bacillus licheniformis*, is more stable at alkaline pH and can be added to detergents to degrade starch-based stains.

(See also Chapter 9, *Biochemistry for advanced biology*.)

Equipment and materials

- water bath set between 20 and 30 °C
- test tubes and rack
- 1 cm^3 graduated pipette
- two 10 cm^3 measuring cylinders
- stop clock
- white tile with sample wells
- dropping pipette
- range of buffers from pH 4.0 to pH 9.0
- Termamyl solution
- starch solution
- iodine solution
- pH paper or pH meter

SAFETY

> Some people are allergic to *enzymes*. Avoid direct contact or inhalation, and clean up spillages immediately.

Experimental procedure

1. Amylase cleaves glycosidic bonds in starch to give a mixture of glucose and short-chain dextrins. Starch gives a blue-black colour with iodine solutions, while the dextrins give a range of brownish purple colours. Do a preliminary experiment to observe this colour change as follows: mix 5 cm^3 starch and 5 cm^3 pH 7 buffer. Place one

drop of iodine into each well of the white tile. Add one drop of the starch mixture to one of the wells and note the characteristic blue-black colour.

2. Now add 1 cm³ enzyme to the starch mixture, shake the test tube and start the stop clock. Take samples every 30 seconds and add to a fresh iodine-containing well each time until no further colour change is seen.
3. Examine the samples in the wells and decide which colour change you will take as your end-point.
4. Now set up one tube for each buffer from the range available. Add 5 cm³ starch and 5 cm³ buffer and incubate for 5 minutes in the water bath, recording the temperature.
5. Add 1 cm³ enzyme solution to the first tube, start the clock and sample every 30 seconds as described in Step 2 above. Record the time taken for the colour to change.
6. Repeat for the other tubes.

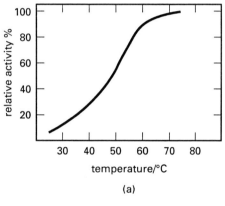

Figure 14.1 Termamyl™ data:
(a) Activity of Termamyl at different temperatures, at pH 9.
(b) Activity of Termamyl of different pH values, at 37 °C.

QUESTIONS
1. Plot the pH profile for your enzyme(s). Comment on the shape of your graph. Estimate the optimum pH for Termamyl.
2. What sources of inaccuracy are there in your experiment?
3. How else could you measure the activity of the enzyme?
4. Look at Figure 14.1, which gives the manufacturer's pH and temperature profiles for Termamyl. Sketch graphs giving the pH profile for three different temperatures, using the same axes.
5. How does the optimum pH of Termamyl compare with that of other enzymes? (See Chapter 9, *Biochemistry for advanced biology*).

© Cambridge University Press 1994

66 EXPERIMENT 14

Teacher/technician notes

Equipment and materials

- water bath set between 20 and 30 °C
- test rubes and rack
- 1 cm^3 graduated pipette
- two 10 cm^3 measuring cylinders
- stop clock
- white tile with sample wells
- dropping pipette
- range of buffers from pH 4.0 to pH 9.0 (see below)
- Termamyl solution (see below)
- starch solution (see below)
- iodine solution (see below)
- pH paper or pH meter (see below)

Range of buffers

To obtain a range of buffers from pH 4.0 to pH 9.0, make up 0.067 M potassium dihydrogenphosphate (9.1 g l^{-1}) and 0.067 M disodium hydrogenphosphate (9.5 g l^{-1}); Table 14.1 gives the composition of buffer mixtures.

Table 14.1 Buffer mixture composition

pH	0.067 M KH$_2$PO$_4$/cm^3	0.067 M Na$_2$HPO$_4$/cm^3
4.5	100	0
5.9	90	10
7.0	40	60
8.0	5	95
8.8	0	100

Termamyl solution From NCBE. It should be stored at 4 °C and diluted 1/500 or 1/1000 in distilled water just before the experiment.

Starch solution A 1 per cent fresh solution made by mixing dry starch to a thin cream in cold water and pouring this into boiling water, boiling briefly and cooling.

Iodine solution Should be 0.001 M (0.254 g l^{-1} in 0.1 per cent potassium iodide solution).

pH paper or pH meter If the pH meter is not available for students, check buffers with pH meter before start of the experiment. They should at least check with pH paper.

Answers to questions

1. See Figure 14.1. The pH optimum is around 7.5.
2. Colour change is complex and somewhat subjective. Timing is only to the nearest 30 seconds.
3. Monitor the appearance of glucose with Benedict's or Fehling's solution.

© Cambridge University Press 1994

4. See Figure 14.2. Three experiments like Experiment 14 have been performed at different temperatures. The activity at all pH values increases with temperature – as suggested by the temperature profile in Figure 14.1.
5. See Table 14.2.

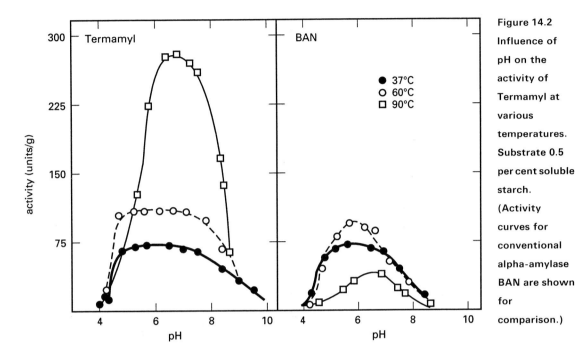

Figure 14.2 Influence of pH on the activity of Termamyl at various temperatures. Substrate 0.5 per cent soluble starch. (Activity curves for conventional alpha-amylase BAN are shown for comparison.)

Table 14.2 The pH maxima of some enzymes

Enzyme	Optimum pH
Pepsin	2.00
Invertase	4.50
Catalase	7.60
Amylase	6.80
Pancreatic lipase	9.00
Urease	7.00

Assessment guide

Experimental skills

1. Works quickly and methodically either alone or in a team (2).
2. Follows procedure correctly with little or no guidance (2).
3. The pH of buffers and temperature of water bath checked and recorded (2).
4. Use of preliminary experiment to assess colour change in a consistent way (2).
5. Sufficient readings (of time) taken (2).

© Cambridge University Press 1994

Experiment 15
The effect of substrate concentration on enzyme action

Aim
To appreciate how the variation of the rate of enzyme-catalysed reactions varies with substrate concentration and link this to the way in which enzymes interact with their substrates.

Introduction
In 1913, Leonor Michaelis and Maud Menten proposed a theory to explain the way enzymes act. Their ideas came from studying the way in which the rate of enzyme-catalysed reactions increases with substrate concentration. They noticed that, at low substrate concentrations, the rate is proportional to concentration, but as concentration increases, the rate levels off until a maximum value is achieved. This rate can only be increased further if more enzyme is added (Figure 15.1).

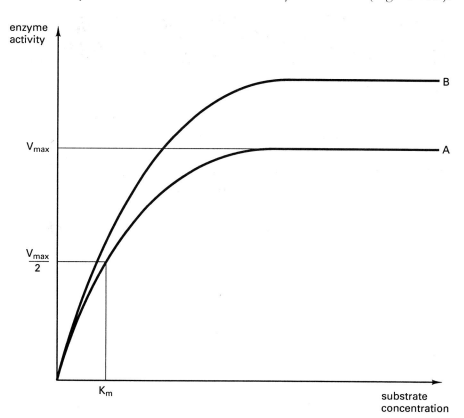

Figure 15.1 The Michaelis–Menten model; in B the enzyme concentration is greater than in A.

Michaelis and Menten defined two important properties from their model, which are used today to compare the properties of different enzymes: K_m, the **Michaelis constant**, is the substrate concentration at which half the **maximum enzyme velocity** (V_{max}) is reached. A low value of K_m means that the enzyme has a high affinity for (binds tightly to) its substrate.

Catalase, which catalyses the breakdown of hydrogen peroxide (H_2O_2) to water and oxygen, is one of the fastest-acting enzymes known; it can break down 200 000 molecules of hydrogen peroxide per second. This action has an important protective action in the cells, for hydrogen peroxide is a toxic by-product of some cellular chemical reactions. Catalase is particularly abundant in the liver.

(See also Chapter 9, *Biochemistry for advanced biology*.)

© Cambridge University Press 1994

Equipment and materials

- mortar and pestle
- muslin filter
- scissors
- 10 cm³ measuring cylinder
- 250 cm³ measuring cylinder
- 250 cm³ beaker
- 250 cm³ side-arm flask
- cork fitted with rubber seal ('Suba-seal')
- 2 cm³ hypodermic syringe and needle
- 100 cm³ gas syringe and connectors
- piece of liver
- fine clean sand
- pH 7 buffer
- 20 volume hydrogen peroxide

SAFETY

The *hydrogen peroxide* solution is corrosive. Use gloves to handle it.

Experimental procedure

1. First make a liver extract by chopping the liver into small pieces and grinding with the sand and about 10 cm³ of buffer.
2. Filter this mixture through the muslin into the small measuring cylinder, transfer to the beaker and make up to about 200 cm³ with buffer. This is the enzyme solution that you will use in these experiments.
3. Transfer 15 cm³ hydrogen peroxide and 10 cm³ buffer to the conical flask.
4. Set up the flask as shown in Figure 15.2.

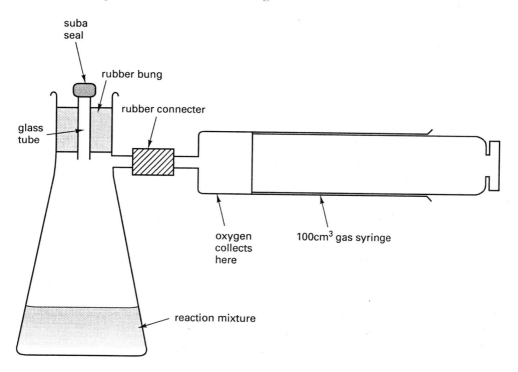

Figure 15.2 Apparatus for Experiment 15.

© Cambridge University Press 1994

5. Make sure the gas syringe plunger is pushed in. Fill the hypodermic syringe with $2\,cm^3$ enzyme extract and add to the side-arm flask through the rubber seal. Shake the flask. Oxygen is produced immediately and pushes out the plunger.
6. Measure the volume of oxygen produced in 15 seconds. If this is not between 40 and $60\,cm^3$, adjust the volume of enzyme added and repeat the experiment.
7. Repeat the experiments with a range of hydrogen peroxide concentrations, obtained by adding 12, 10, 8, 6, 4, 2 and $0\,cm^3$ to buffer, always keeping the total volume in the flask at $25\,cm^3$.
8. Carry out a control experiment with no enzyme to see if there is any breakdown of hydrogen peroxide. If gas is produced, subtract the volume produced in 15 seconds from the above readings.

QUESTIONS

1. Plot the rate (volume of oxygen produced divided by time) against concentration (20 volume hydrogen peroxide is 1.74 M).
2. Explain the shape of your graph. Do the results support the Michaelis–Menten model? If so, record V_{max} and K_m.
3. Sketch the graph you would expect if you doubled the enzyme concentration.
4. Why is it important to work with buffered solutions?
5. Which factors are kept constant in this experiment? What could you do to make sure they remain constant?

Teacher/technician notes

Equipment and materials

- mortar and pestle
- muslin filter
- scissors
- 10 cm^3 measuring cylinders (three per student or group)
- 250 cm^3 measuring cylinder
- 250 cm^3 beaker
- 250 cm^3 side-arm flask
- cork fitted with 'Suba-seal'
- 2 cm^3 hypodermic syringe and needle
- 100 cm^3 gas syringe and connectors
- piece of liver
- fine clean sand
- pH 7 buffer (recipe as in Experiment 3)
- 20 volume hydrogen peroxide (keep this refrigerated, away from bright light)

Answers to questions

1. —
2. The graph should obey the Michaelis–Menten model. K_m for catalase is 1.0 M.
3. It has a greater V_{max} but the same K_m and the same overall shape.
4. Buffers keep the pH constant.
5. Temperature should be constant too – ensure this by working in a thermostatted water bath.

Assessment guide

Observation and recording

1. Makes correct number of readings of gas volume in a consistent way (all at 15 seconds) (2).
2. Pushes plunger in to 'zero' the syringe before each experiment (2).
3. Checks and adjusts gas volume to between 40 and 60 cm^3 after first experiment (2).
4. Volume kept constant at 25 cm^3 (2).
5. Results recorded in an appropriate format (2).

© Cambridge University Press 1994

Experiment 16
Designing a washing powder

Aim
To investigate enzymes that degrade biological stains and to evaluate different mixtures of these enzymes in washing powders.

Introduction
So-called 'biological' detergents account for 85 per cent of the Western European washing powder market. These contain enzymes that degrade biological stains such as blood, egg and milk. The original biological detergents, launched in the 1960s, contained proteases to break down protein-based stains. Then, in the late 1980s, **amylase** and **lipase** were added to break down starch and fat-based stains. Recently, **cellulases** have also been added. These condition cotton-based fabrics by breaking down damaged cellulose fibres, which roughen the surface of the fabric after repeated washing.

Temperature is an important factor in determining the effectiveness of a wash. Protein stains are harder to remove at temperatures above 45 °C, while starchy and greasy stains are only removed at high temperatures. Enzymes help to overcome this problem. A powder containing an enzyme mixture allows effective washing at lower temperatures, decreasing energy consumption.

(See also Chapter 9, *Biochemistry for advanced biology*.)

Equipment and materials
- 100 cm^3 beakers
- water bath
- dropping pipettes
- set of detergent enzymes – this includes a high- and low-temperature protease, a lipase, an amylase and a cellulase
- biological detergent
- non-biological detergent
- small squares of white cotton to fit beakers
- range of food stains, e.g. milk, gravy powder, baby food, tinned spaghetti, tea

SAFETY

Do not use blood as a protein stain in this experiment.

Some people are allergic to *enzymes* – use rubber gloves when handling the detergent.

Experimental procedure

1. Plan your investigation. Consider the following.
   (a) How to obtain a range of stains. You should include protein, starch and fat-based stains alone and in mixtures.
   (b) The evaluation of the effects of the individual enzymes (these enzymes need to be no more than 1 per cent of the mixture, and less is usually effective). Consider the temperatures at which you

will test, and how you will evaluate the effectiveness of each detergent mixture (consult the data sheets of Figure 16.1).

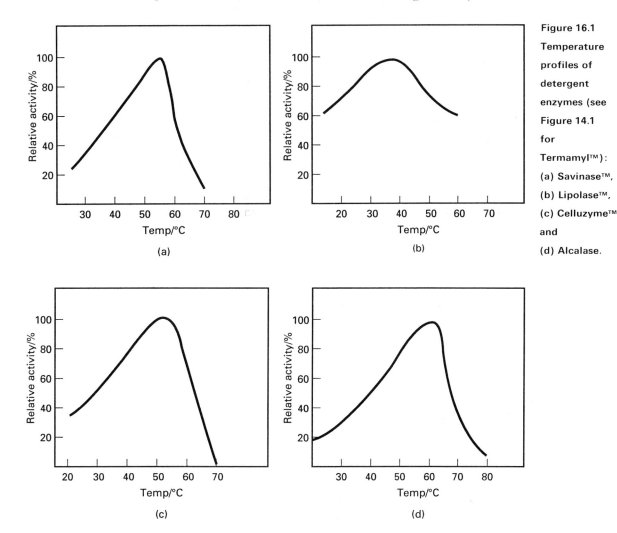

Figure 16.1 Temperature profiles of detergent enzymes (see Figure 14.1 for Termamyl™): (a) Savinase™, (b) Lipolase™, (c) Celluzyme™ and (d) Alcalase.

(c) The formulation of washing powder mixtures based on your observations.
(d) The inclusion of controls and presentation of your results.
2. Check your plan with your teacher.
3. Carry out the investigation.

QUESTIONS

1. How does the performance of your detergent compare with that of the commercial brand?
2. What was the most effective mixture you tested? Write instructions and advertising for a packet of your detergent.
3. How would you design a detergent mixture to appeal to (a) parents with a large load of baby clothes to wash, (b) someone who is about to throw away a favourite blouse or shirt because the surface has become roughened by frequent washing and (c) someone who travels a lot?
4. Why could a detergent lipase be ineffective in a mixture containing a protease?
5. Why are protein stains hard to remove above 45 °C in ordinary detergent?
6. Outline how bulk quantities of these detergent enzymes might be obtained.

© Cambridge University Press 1994

Teacher/technician notes

Equipment and materials

- $100\,cm^3$ beakers
- water bath
- dropping pipettes (to apply stains)
- set of detergent enzymes (see below)
- biological detergent (as a comparison)
- non-biological detergent (as a control)
- small squares of white cotton to fit beakers (students should stain these in pairs for 'before' and 'after' comparisons)
- range of food stains, e.g. egg (fat and protein), milk (fat and protein), gravy powder (starch and protein), baby food (starchy), tinned spaghetti (starchy); try to have packets/tins available as this gives clear information about starch/protein/fat content or encourage students to bring in their own choices

Set of detergent enzymes This includes a high- and low-temperature protease, a lipase, an amylase and a cellulase (available in 50 g quantities from the National Centre for Biotechnology Education). These are powders that can be used alone as a 0.1 per cent solution in distilled water or as a 1–2 per cent mixture with non-biological powder and then diluted to a 5 per cent solution in distilled water. At these concentrations and at a temperature of 40–50 °C these will remove most biological stains.

A reasonable plan would be to investigate the effects of the enzymes on a range of stains separately, and then make up enzyme mixtures. These can be tested alone and then with non-biological detergents. It may be more practicable for each student in a group to develop a detergent mixture for one kind of stain and have the group share their results. Comparison of performance with commercial biological detergent should be made, and temperature and washing time should be noted and, if time permits, investigated for optimum performance.

Answers to questions

1. Refer to instructions on detergent packet to compare temperature, detergent concentration and wash times for rough comparison. Enzyme mixtures alone will be less effective because there is no soap, water softener or surfactant – the stains may tend to re-adhere to the fabric.
2. This should include range of stains removed, temperature, amount of detergent used per volume of water, time of wash and need to avoid direct skin contact with enzymes.
3. (a) Amylase will remove starchy baby food, lipase will remove milk stains and protease will remove egg stains. Emphasise the importance of thorough rinsing of wash, so that enzyme traces do not irritate baby's skin. (b) Include cellulase as a conditioner and use mixture of enzymes that allows low-temperature wash. (c) A fairly concentrated mixture of protease, amylase and lipase to work in cold water is

needed in case no hot water is available in cheap hotel rooms or occasions where a quick wash is necessary.

4. The protease could break down the lipase. Experiments during the development of the detergent would select a compatible mixture.
5. Denaturation decreases their solubility in water.
6. Detergent enzymes come from bacteria and fungi, so you would grow large numbers in a fermenter and extract the enzymes from them. Some enzymes are secreted into the fermentation medium. Those which are not are released from inside the cells by breaking up the cells with a machine called a homogeniser. Then the medium is centrifuged to get rid of cell debris, and the enzyme is precipitated out of the supernatant.

Assessment guide

Experimental design

1. Logical and detailed work plan, delegating among group if necessary (2).
2. Consideration given to concentration, temperature and wash time of enzyme/detergent mixtures (2).
3. Range of stains including protein, starch and fat alone plus mixtures mentioned (2).
4. Method for evaluating performance and presenting results (2).
5. Controls of water, non-biological and biological detergents included (2).

© Cambridge University Press 1994

Experiment 17
Investigating enzyme specificity using molecular modelling

Aim
This investigation helps you to appreciate the three-dimensional structures of proteins and how small changes in amino acid sequence can lead to significant alterations in specificity.

Introduction
X-ray crystallography has shown that the overall shapes of the digestive enzymes chymotrypsin, trypsin and elastase are very similar. Further experiments have shown that the active sites of these enzymes are identical, consisting of a **catalytic triad** of aspartic acid, histidine and serine. Because the serine residue is the key to the catalytic action, which cleaves peptide bonds in proteins, these three enzymes are known as **serine proteases**. The serine protease family also includes the proteins responsible for the clotting of blood, such as thrombin.

The serine proteases are all highly specific in their choice of substrate, despite their overall similarity in structure. **Chymotrypsin** cleaves polypeptide chains at tryptophan, phenylalanine and tyrosine residues, **trypsin** at arginine and lysine residues, while **elastase** is specific for small residues such as glycine and alanine.

The three enzymes all have a set of amino acids near the active site known as the **binding pocket**. These residues are far apart in the polypeptide chain, but when the protein folds up into its tertiary structure they are brought close together. The binding pocket holds onto the side chain of the residues specified above, while the active site cleaves the polypeptide chain. So, in chymotrypsin, the side chain of a tryptophan residue slips into the binding pocket while the peptide bond linking the tryptophan to its neighbour is broken at the active site.

The amino acids that line a binding pocket determine its size and shape, and this gives the serine proteases their specificity. It is possible to design new protease enzymes with different specificities by altering the amino acid composition of the binding pocket. This is **protein engineering**. A gene for the new protein is made chemically, by substituting the codon for the altered amino acid for the one in the native protein. The artificial gene is then cloned into cells, which synthesise it.

(See also Chapter 2, *Biochemistry for advanced biology*.)

Equipment and materials
- photocopied pictures of trypsin and of the binding pockets of chymotrypsin, trypsin and elastase (Figure 17.1)
- chart of amino acid structures (Figure 17.2)
- molecular modelling kit
- desktop molecular modeller (optional)
- wire
- Blu-Tack or plasticine

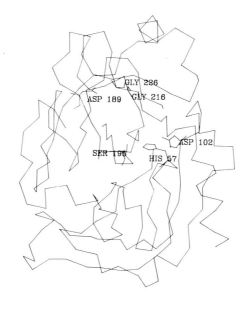

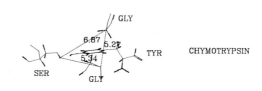

Figure 17.1
(a) Trypsin molecule with binding pocket and catalytic triad residues. (b) Binding pockets of serine proteases. Dotted lines give outline and dimensions (in angstroms, one angstrom = 0.1 nm) for chymotrypsin binding pocket. The dimensions for the binding pocket are the same for all three enzymes although they are shown only for chymotrypsin. The distances marked are between the alpha carbon atoms of the labelled amino acids.

Procedure
Read through and answer the following questions.

QUESTIONS
1. Look at the picture of the trypsin molecule (whole protein) and circle the binding pocket and the catalytic triad.
2. Draw a line diagram of the primary structure of trypsin, marking the residues of the binding pocket and the catalytic triad. Is there an obvious relationship between primary and tertiary structure?
3. Now look at the picture of the binding pocket of chymotrypsin. Only the amino acids of the binding pocket that give the enzyme its specificity are displayed. Two glycines 'guard' the entrance to the pocket, while a serine 'sits' at the base. The dimensions of the pocket are given in angstroms (a unit that is often used in molecular modelling – one angstrom is 0.1 nm). These dimensions come from the known three-dimensional structure of

© Cambridge University Press 1994

chymotrypsin. Mark the side chain of the amino acid of the substrate (tyrosine). Check the amino acid chart and explain the specificity of chymotrypsin.

Figure 17.2 Amino acids (N = non-polar, P = polar, A = acidic, B = basic).

Basic structure:

$$\begin{array}{c} COOH \\ | \\ H-C-NH_2 \\ | \\ R \end{array}$$

Alanine, Ala (N)	$R = -CH_3$
Arginine, Arg (B)	$R = -(CH_2)_3-NH-C(\!\!\begin{array}{c}{=}NH\\ {\backslash}NH_2\end{array}\!\!)$
Asparagine, Asn (P)	$R = -CH_2-CO-NH_2$
Aspartic acid, Asp (A)	$R = -CH_2-COOH$
Cysteine, Cys (P)	$R = -CH_2-SH$
Glutamine, Gln (P)	$R = -(CH_2)_2-CO-NH_2$
Glutamic acid, Glu (A)	$R = -(CH_2)_2-COOH$
Glycine, Gly (N)	$R = -H$
Histidine, His (B)	$R = -CH_2-\text{(imidazole ring)}$
Isoleucine, Ile (N)	$R = -C(H)(CH_3)-C_2H_5$
Leucine, Leu (N)	$R = -CH_2-CH(CH_3)_2$
Lysine, Lys (B)	$R = -(CH_2)_4-NH_2$
Methionine, Met (N)	$R = -(CH_2)_2-S-CH_3$
Phenylalanine, Phe (N)	$R = -CH_2-\text{(phenyl)}$
Proline, Pro (N)	(structure of whole molecule: ring with $CH_2-CH_2-CH_2-NH-CH-COOH$)
Serine, Ser (P)	$R = -CH_2-OH$
Threonine, Thr (P)	$R = -C(H)(OH)-CH_3$
Tryptophan, Trp (N)	$R = -CH_2-\text{(indole)}$
Tyrosine, Tyr (P)	$R = -CH_2-\text{(phenyl)}-OH$
Valine, Val (N)	$R = -CH(CH_3)_2$

4. Confirm your conclusion by making a model of the binding pocket. First construct the three amino acids that are shown in Figure 17.1 (glycine, glycine and serine at the sides and base of the binding pocket respectively), then arrange them according to the dimensions given (fix with wire and Blu-Tack or plasticine) to make a binding pocket. If you have the desktop molecular modeller, simulate Figure 17.1 on the screen. Now make some amino acid substrates (remembering that these would be part of a polypeptide chain) and try to fit them into the binding pocket of chymotrypsin. Which fit comfortably? Which ones 'rattle' (and will diffuse away). Are there any that are too big to enter? (Remember, it is only the side chain that goes into the binding pocket.)

5. Look at the pictures of the elastase and trypsin binding pockets. List the differences in the residues 'guarding the entrance' and 'sitting at the base' of the pocket. Then repeat questions 3 and 4 above for these proteins.

6. Suggest a way of altering the trypsin binding pocket to make it specific for acidic rather than basic amino acids.
7. Suggest a way of making elastase less specific.
8. For questions 6 and 7, write down the codon change that would be necessary in the gene for the new proteins.

Teacher notes

Equipment and materials

- computer-generated pictures of trypsin, binding pockets of chymotrypsin, trypsin and elastase based on X-ray crystallographic models (Figure 17.1, one photocopy for each student)
- molecular modelling kit (e.g. Philip Harris, Biochemistry Set)
- chart of amino acid structures (Figure 17.2)
- desktop molecular modeller from Oxford University Press (works with IBM compatibles)

Answers to questions

1. Asp 189, Gly 226 and Gly 216 are the binding pocket residues; Ser 195, Asp 102, His 57 is the catalytic triad.
2. The above residues come in numerical order in the primary sequence, so there is no obvious relationship between primary and tertiary structure (worth mentioning that this is a key problem in molecular biology at the current time).
3–5. The residues 'guarding the entrance' of the chymotrypsin binding pocket are glycines, the smallest possible. So they let in the biggest residues – tryptophan, tyrosine, phenylalanine. Smaller residues diffuse in, 'rattle around' and diffuse out again – nothing happens. The valine and threonine residues 'guarding the entrance' of the elastase binding pocket block the entry of all but the smallest residues such as alanine and glycine. Trypsin has an acidic residue – aspartic acid – sitting at the base of its binding pocket. This specifically attracts in basic residues, i.e. lysine and arginine. Elastase and chymotrypsin do not have an acidic amino acid at the bottom of the binding pocket.
6. Change the Asp 189 to Lys or Arg, which will bind acid side chains.
7. Substitute Gly for the Val and Thr.
8. AAA (Lys) or AGA (Arg) for GAU (Asp), and GGU (Gly) for GUU (Val) and ACU (Thr).

Assessment guide

Data handling

1. Mapping of binding pocket and catalytic triad onto primary structure diagram (2).
2. Translation of data in binding pocket diagrams into models (2).
3. Listing differences in amino acids lining the binding pockets of serine proteases (2).
4. Explanation of specificity of serine proteases in terms of diagrams and models showing differences in residues in binding pockets (2).
5. Extension of conclusions above to protein engineering strategy to alter enzyme specificity (2).

© Cambridge University Press 1994

Experiment 18
Chemicals and cancer (a data-handling exercise)

Aim
This investigation gives you practice in handling data and simple statistics. It also gives you a chance to appreciate how the data obtained in research are used to draw conclusions that can influence public health policy. Note that this exercise is rather different from the classroom experiments you are used to. You may even find it quite demanding. It is included as an example of how scientific data are actually obtained and applied in real life.

Introduction
Epidemiology involves investigating the effects of external influences, such as infection, diet and toxic substances, on the incidence of disease. Most cancers are generally thought to be caused by smoking, diet and genetic factors. Some chemicals (carcinogens) can cause cancer. Epidemiological studies of cancer in industrial workers have produced useful information about carcinogens that has been used in drawing up safety guidelines.

Dioxin is a chlorine-containing organic compound that is widely distributed in the environment because it is a frequent contaminant of herbicides, as well as being produced by paper making and incineration.

In some animals, dioxin is a potent carcinogen, and there is concern about its effects in humans. Studies of workers exposed seem to suggest that it may cause human cancer.

You will be looking at a recent study, published in a medical journal (*The Lancet*), that is concerned with the incidence of cancer in people who have worked with chemicals contaminated with dioxin.

(See also Chapter 7, *Biochemistry for advanced biology*.)

Equipment
- calculator
- graph paper

Procedure
Extracts and data from the paper in *The Lancet* are presented below. You should work through these, answering the questions as they occur. Note that a glossary follows the extract – you might find it helpful to refer to this as you read.

A summary of the paper

Cancer mortality among workers in chemical plant contaminated with dioxin
A. Manz and co-workers, *The Lancet*, 19 October 1991

Dioxin can arise as a contaminant in the production of herbicides. It causes chloracne in those exposed to it, but its human carcinogenicity has been a matter of dispute. We report here a mortality follow-up of 1583 workers (1184 men, 399 women) employed in a chemical plant in Germany that produced herbicides, including products

© Cambridge University Press 1994

contaminated with dioxin. Production of dioxin was reduced from 1954 after an outbreak of chloracne.

Vital status up to 1989 was determined for 97.1 per cent of workers hired between 1952 and 1984, and 367 deaths (313 men, 54 women) were recorded. A malignant neoplasm was the underlying cause of death in 93 men and 20 women. Standardised mortality ratios (SMR) were calculated with, as references, national mortality statistics for West Germany and deaths in a cohort of male gas workers; for total cancer mortality they were 1.24 (95 per cent confidence interval 1.00–1.52) and 1.39 (1.10–1.75) respectively among men. Cancer mortality was increased among men with 20 or more years of employment (SMR = 1.87 compared to Germany as a whole and 1.82 compared to gas workers) and among men who began employment before 1955 (SMR = 1.61 and 1.87). The group with the highest exposure to dioxin had SMRs of 1.42 and 1.78. Only 7 per cent of cohort women worked in the high-exposure locations in the plant, compared with 39.6 per cent of men, and no (overall) increased risk of cancer mortality was observed among women; but breast cancer mortality was raised (SMR = 2.15).

These results, together with a US occupational study and a German investigation of accidental exposure, support the hypothesis that dioxin is a human carcinogen.

Glossary

- **chloracne**   a severe skin condition linked with dioxin exposure
- **carcinogenicity**   ability to cause cancer
- **vital status**   whether alive or dead
- **malignant neoplasm**   clinical term for cancer
- **standardised mortality ratio**   a comparison of actual deaths in a particular population compared with those expected in a control population
- **cohort**   individuals of the same age
- **95 per cent confidence interval**   there is a 95 per cent chance that the value lies within this range
- **hypothesis**   a suggested explanation for experimental observations that forms the basis for further investigations
- **null hypothesis** (see question 3)   the assumption that there is no significant difference between experimental and control groups in an experiment; if the null hypothesis is disproved by the data, then the experimental hypothesis is proved

QUESTIONS

1. What specific question was being asked by this investigation?
2. Why do you think two different control groups were used?
3. What would the null hypothesis be for this investigation?

The data

The researchers used medical records to find out the cause of death in the chemical and gas workers, and death certificates to find out the cause of death in the general population. They used working patterns to determine exposure to dioxin and divided the workers into three groups on this basis – high (group 1), medium (group 2) and low exposure (group 3). To see how accurate these groupings were, they took direct readings of dioxin concentration in the fatty tissue of 48 volunteers from the groups of workers being studied (see Table 18.1).

Table 18.1 Dioxin levels in tissue samples

Group 1	
Number of subjects tested	37
Mean dioxin level	296 ng kg^{-1}
Standard deviation	479 ng kg^{-1}
Median	137 ng kg^{-1}
Groups 2 and 3 combined	
Number of subjects tested	11
Mean dioxin level	83 ng kg^{-1}
Standard deviation	73 ng kg^{-1}
Median	60 ng kg^{-1}
Normal population	
Dioxin levels between 7 and 20 ng kg^{-1}	

*One nanogram (ng) is 10^{-9} g.

Note the meaning of the following terms before starting question 4.

- **mean (m)** the sum of all observations made (here, the observations are dioxin levels) divided by the number of observations (here, the number of subjects)
- **median** the 'middle' value of a set of observations when they are arranged in increasing order
- **standard deviation (s)** a measure of the 'spread' of a set of data about the mean, given by the formula:

$$s = \sqrt{\left(\frac{\sum (x - m)^2}{N}\right)}$$

where the x values are the observations (and $x - m$ is the difference between an observation and the mean) and N is the total number of observations

- **distribution curve** a graph showing observations (here, the dioxin levels) on the x axis and the frequency of those observations in an experiment (here, the number of subjects with a particular dioxin level) on the y axis

QUESTIONS

4. Draw a possible distribution curve for the dioxin levels in both groups of workers, marking the mean, the median and the standard deviation. (Note: you do not need the above formula to do this.)
5. Why do you think the researchers did not measure dioxin levels in all their subjects?
6. In what ways does the type of data used here differ from those gathered in experiments you may have done on living populations? What factors determine the quality of these data?

84 EXPERIMENT 18

Mortality data

The mortality data are summarised in Tables 18.2 and 18.3.

Table 18.2 Vital status of all exposed groups

Group	Number
Men	
Alive	836
Dead	313
Unknown	35
Women	
Alive	334
Dead	54
Unknown	11

Table 18.3 Causes of death in the exposed groups

Group	Number
Men	
Malignant neoplasms	93
Other natural causes	172
Accident	20
Suicide	22
Other	1
Unknown	5
Women	
Malignant neoplasms	20
Other natural causes	28
Accident	2
Suicide	2
Other	0
Unknown	2

QUESTION 7. Convert the mortality data of Tables 18.2 and 18.3 into percentages and present them diagrammatically.

Standard mortality ratio (Table 18.4) is the number of deaths observed divided by the number of deaths expected. If, for example, the SMR had a value of 1.5, this would mean that the number of deaths observed was one and a half times, or 50 per cent above, the number expected.

QUESTIONS 8. Fill in the SMR values in Table 18.4, using the definition given above. How do you account for the differences?

9. What percentage of the control group would have died of cancer by the time of this study, according to these figures? List some possible causes of cancer in such a control group.

© Cambridge University Press 1994

Table 18.4 Standardised mortality ratios (SMRs) for malignant neoplasms, compared to the West German population

Cohort (number)	Observed deaths	Expected deaths	SMR
Total (1148)	93	75	
Entry before 1955 (296)	51	31.6	
Duration of employment more than 20 years (118)	18	9.7	

Further information

Group 2 was also exposed to benzene. Group 3 was exposed to dimethyl sulphate. Both of these chemicals are known to cause cancer. A total of 73 per cent of the chemical workers said they smoked, as did 76 per cent of the gas worker control subjects. So it could be argued that at least some of the chemical workers were exposed to other influences that could have caused the cancers.

The following month, a letter criticising this research was published by *The Lancet*. The following three points were made in the letter. (a) Medical records contain more detailed information than death certificates. This could have led to an overestimate of the SMRs. (b) The numbers of cancer deaths are too small to form a reliable conclusion about a link between dioxin and cancer. (c) The cancers could have been caused by smoking.

QUESTION

10. Write a short summary of this research, as if for a newspaper or magazine for the general public, making suggestions for any action that could follow from it.

Teacher notes

Equipment

- calculator
- graph paper

Dioxin is the common name for 2,3,7,8-tetrachlorodibenzodioxin (TCDD), one of a family of chemicals known generally as dioxins. The common name has been used throughout this exercise. TCDD is the most toxic of the dioxins, according to animal studies.

The data presented here are necessarily selective, but cover the main points in the abstract. Data for females are not included in Table 18.4 because the numbers are so small. Data for the gas workers were also excluded to simplify the exercise. Only the conclusions are retained.

Answers to questions

1. The question being asked was: 'Does dioxin exposure increase the incidence of cancer in humans?'
2. Chemical workers are exposed to many potential toxic chemicals as well as dioxin. A comparison with just the general population might be showing up effects from this background exposure to chemicals. A second comparison, to gas workers who are exposed to chemicals, but not dioxin, is a stronger control than the general population.
3. The null hypothesis would be: 'Dioxin exposure does not increase the incidence of cancer in humans.'

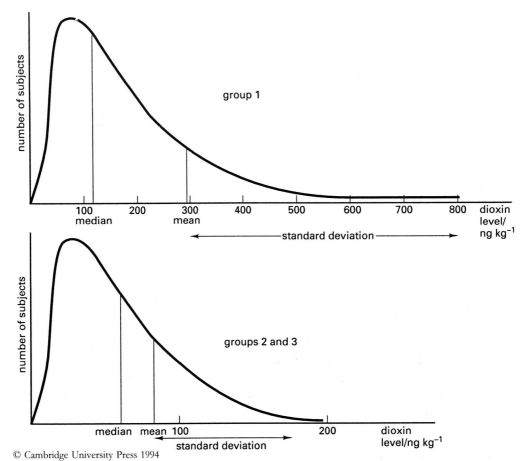

Figure 18.1 Dioxin levels in chemical workers.

© Cambridge University Press 1994

4. See Figure 18.1. The graphs should have a positive skew. Mode, median and mean do not coincide. Standard deviation is a measure of the spread of the data and it is given by:

$$s = \sqrt{\left(\frac{\sum (x - m)^2}{N}\right)}$$

The standard deviations in the dioxin levels are large – there are some subjects with very high levels, more than with very low levels.

5. Measuring levels of dioxin in fatty tissue is clinically harder than taking a blood sample and more unpleasant for volunteers. The researchers would have needed to have tracked down all the subjects, which is more expensive and time-consuming than looking at their work records. Here we are looking at a sample within a sample, to verify that the work records are giving a reasonable measure of actual dioxin exposure.

6. The sample size is larger than any that students are likely to have dealt with in their experiments and it is not based on direct observation. The accuracy of the medical and work records determines the quality of the data. These have been kept over a period of time and were not the work of the researchers themselves. This automatically raises the question of accuracy. There are no readings that can be repeated to check reproducibility.

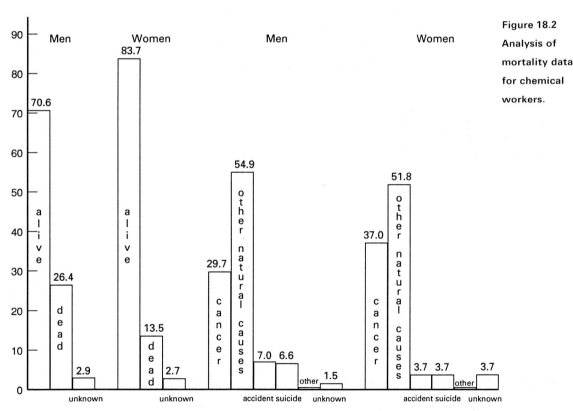

Figure 18.2 Analysis of mortality data for chemical workers.

7. See Figure 18.2. Bar charts and pictograms also acceptable.
8. SMR values are 1.24, 1.61 and 1.86. Production of dioxin decreased after 1955, so workers entering after this date had lower exposure; also the fewer years worked, the lower the exposure. Data support the hypothesis that dioxin exposure increases the incidence of cancer in humans.

9. Out of 1148 control subjects, 75 would be expected to have died of cancer by the time of the study. This is 6.5 per cent. Obviously as time goes by this would increase, to a maximum of about 25 per cent – this can be seen by checking the percentages for men who entered before 1955 (10.7 per cent) and those with more than 20 years employment (8.2 per cent). These men would tend to be older and as a population ages the numbers dying of cancer increase. Causes of cancer among the general population include smoking, diet and genetic disposition (emphasise that most scientists agree that smoking is the leading cause of cancer).

10. This should mention the main conclusions of the study but point out some of its limitations – difficulty of controlling for smoking and other chemical exposures, the long time periods, the reliance on records that could be inaccurate. However, other studies (see references in the quoted paper in *The Lancet*) also suggest that dioxin is a carcinogen. Follow-up studies should be carried out on industrial workers and the general population.

Assessment guide

Data handling

1. Correct calculations (questions 7, 8 and 9) (2).
2. Presentation of data (question 7) (2).
3. Valid conclusions drawn (questions 9 and 10) (2).
4. Understanding methodology (questions 1, 2, 3, 5 and 6) (2).
5. Use of statistics (questions 3 and 4) (2).

List of suppliers

Philip Harris Education
Lynn Lane
Shenstone
Lichfield
Staffordshire WS14 0EE
01543–480077
(for general laboratory supplies)

Philip Harris Education
2 North Avenue
Clydebank Business Park
Clydebank
Glasgow GS1 2DR
0141–952 9538
(for general laboratory supplies)

Philip Harris Biological
Oldmixon
Weston-super-Mare
Avon BS24 9BJ
01934–413063
(for biological supplies)

Aldrich Chemical Company Ltd
The Old Brickyard
New Road
Gillingham
Dorset SP8 4JL
01747–824414
(for chemical supplies)

National Centre for Biotechnology Education
Department of Microbiology
University of Reading
Whiteknights
PO Box 228
Reading RG6 2AJ
01734–873743
(enzymes)

Oxford University Press
Walton Street
Oxford OX2 6DP
01865–56767
(desktop molecular modeller)

Bibliography

N. P. O. Green, G. W. Stout and D. J. Taylor, *Biological Science*, vols 1 and 2 (Cambridge, Cambridge University Press, 1990)

McCance and Widdowson, *The Composition of Foods*, 5th edn (London, Ministry of Agriculture, Fisheries and Food/Royal Society of Chemistry)

A. Manz *et al.*, Cancer mortality among workers in chemical plant contaminated with dioxin, *Lancet*, **303**, 19 October 1991

E. K. Watson *et al.*, Screening for carriers of cystic fibrosis through primary health care services, *Lancet*, **303**, 31 August 1991, pp. 504–7

D. J. Weatherall, *The New Genetics and Clinical Practice* (Oxford, Oxford University Press, 1991)

The *Newsletter of the National Centre for Biotechnology Education* (free with membership of the Schools' Biotechnology Club) is a useful source of tips and background material, as are the *School Science Review* and the *Journal of Biological Education*

Index

C_3 plants, 30
C_4 plants, 30
callus, 9, 12
cancer, case study of, 81–8
carcinogens, *see* cancer
cells, 1–8
   animal, 1–3
   bacterial, 1–3
   plant, 1–4
chromatography, high-performance liquid and paper, 14–15
clones, 9

detergents, 72–5
DNA
   denaturation of, 20
   extraction of, 20–2
DNA fingerprinting, 44–51
DNA testing, 44–51

enzymes, 52–6, 57–9, 72–5
   amylase, 72–5
   catalase, 68–71
   cellulase, 72–5
   chymosin, 60–3
   chymotrypsin, 76–9
   effect of substrate concentration on activity of, 68–71
   elastase, 76–9
   lipase, 72–5
   molecular modelling of, 76–8
   pepsin, 57–9, 64
   pH dependence of, 64–7
   specificity of, 76–8
   thermal stability of, 58–61, 60–3
   trypsin, 76–8
epidemiology, 81

food, nutrients in, 23–6
food testing, 23–6

genetic disease(s), 41, 44–8, 51
   cystic fibrosis, 47–8
   haemophilia, 45–7
   Huntington's chorea, 41, 48
   motor neurone disease, 41

haemoglobin, 14
high-performance liquid chromatography, 14–15
human genome project, 41–3

paper chromatography, 14–19, 27–31
photosynthesis, 32–6
   effect of sunlight on, 42–6
plant hormones, 9–13
plant tissue culture, 9–13
polymorphism, 44–5
protein engineering, 76, 79, 80
proteins, 14
   primary structure of, 14
   sequence analysis of, 14–19

respiration, 37–40
   aerobic, 37
   anaerobic, 37
restriction fragment length polymorphisms (RFLP), 45–6, 48

Southern blotting, 44–5

totipotency, 9